# Mast Cells and Tumours

Domenico Ribatti • Enrico Crivellato

# Mast Cells and Tumours

## From Biology to Clinic

Springer

Prof. Domenico Ribatti
Department of Human Anatomy and Histology
University of Bari Medical School
Policlinico
Piazza Giulio Cesare 11
70124 Bari
Italy
ribatti@anatomia.uniba.it

Prof. Enrico Crivellato
Section of Anatomy
Department of Experimental and Clinical Medicine
University of Udine Medical School
Piazzale Kolbe 3
33100 Udine
Italy
enrico.crivellato@uniud.it

ISBN 978-94-007-1468-7 e-ISBN 978-94-007-1469-4
DOI 10.1007/978-94-007-1469-4
Springer Dordrecht Heidelberg London New York

Library of Congress Control Number: 2011927938



Printed on acid-free paper

Springer is part of Springer Science+Business Media (www.springer.com)

*vom physiologischen dürfte man sie vorläufig als*
*Mastzellen bezeichn*
(from Paul Ehrlich's doctoral thesis, 1878)

# Contents

# Chapter 1
# Introduction

The mast cell (MC) is a type of bone marrow-derived, tissue-homing secretory cell, which was identified by Paul Ehrlich in 1878 and first described in his doctoral thesis. In normal histological sections, the MC shows a very characteristic picture which consists in an abundant cytoplasm filled with thousands of granules. It is a ubiquitous cell which populates the connective tissues of the different organs and locates in particular to the lying surfaces of the body, such as the skin and the gastrointestinal and respiratory tracts. Inner sites, however, as the central nervous system, are also populated by MCs.

MCs have long been considered to be primitive cells, perhaps the surviving remnant of an ancient model of defensive system. These extant cells have been regarded to be mainly able to produce harmful responses to otherwise not threatening inhaled, ingested, or injected antigens. Indeed, MCs have long been implicated in the pathogenesis of allergic reactions although they have also been recognized to participate to certain protective responses to parasites. Recently, however, our perception of MCs has dramatically changed mostly on the base of studies in rodents. These studies have contributed to forge a novel view of MCs. These cells are currently conceived as tissue elements operating as "sentinels" and "effectors" in different defensive settings, being capable to orchestrate inflammatory responses and providing a series of immunoregulatory strategies in the context of both innate and adaptive immunity. On the one hand, they are competent in defending the host against multiple microbiological challenges as well as against the venom of bees and snakes. On the other hand, they are endowed with the capacity to modulate antigen presentation and lymphocyte recruitment to draining lymph nodes as well as up- or down-modulate immune responses. Besides defensive functions, MCs are capable to express functional properties which can be reconciled with the concept of "tissue homeostasis". Indeed, they contribute to organize angiogenic responses and to protect tissue damaging by arranging tissue repair processes after injury. By contrast, MCs are currently regarded as crucial elements in initiating or favouring or worsening a series of pathological conditions such as autoimmunity, coronary spasms, cardiomyopathy, atherosclerosis, myocardial ischemia and aortic aneurysms.

D. Ribatti, E. Crivellato, *Mast Cells and Tumours*,
DOI 10.1007/978-94-007-1469-4_1, © Springer Science+Business Media B.V. 2011

The possible connection of MCs with tumour growth and progress was early adumbrated by Paul Ehrlich himself and by his pupil Westphal. Indeed, most tumours contain inflammatory cell infiltrates, which often include plentiful MCs. Thus, the importance of a potential functional link between chronic inflammation and cancer has long been recognized and the question as to the possible contribution of MCs to tumour development has progressively been emerged. A major point linking MCs to cancer is the well recognized capacity of these cells to synthesize and release potent angiogenic compounds. Beside angiogenic signals, however, MCs synthesize and release a number of mediators capable to influence tumour cell growth directly or indirectly and the role played by MCs in tumour development is complex and controversial. Although some evidence suggests that MCs can promote tumorigenesis and tumour progression, there are clinical conditions as well as experimental models in which these cells seem to have functions that favour the host. Thus, the involvement of MCs in tumour onset and tumour progress is multifaceted and far from being settled.

This book focuses on the relationship between MCs and tumour cells, in particular on the impact of MCs on tumour cell biology. Much has been investigated in the past on this issue but a series of recent studies has greatly contributed to clarify some important aspects regarding the specific involvement of MCs in the tumour microenvironment and the tumour fate, with particular emphasis on the capacity of these cells to stimulate tumour growth by promoting angiogenesis and lymphangiogenesis. This book is divided into eight chapters. In the first two chapters, after an introductory historical prologoue, the general biology of MCs, their multifarious functional activities and involvement in health and disease, as well as their composite mediator profile will be revised. Next, the reader will be introduced into a chapter devoted to the general features of tumour cells, the distinctive characteristics of the tumour microenvironment and Paget's "seed and soil" theory. The special property of the angiogenic process which accompanies tumour growth is discussed in the fourth chapter. The pro and contra MCs involvement in tumour growth in humans is the theme of the subsequent chapter. Then, the cross-talk between MCs and tumour cells will tentatively be dissected and evaluated in the context of the signalling networks operating within the neoplastic scenario, with special emphasis to the processes of MC-induced angiogenesis and lymphangiogenesis. *In vitro* studies as well as investigations conducted in different experimental and clinical settings will be taken into account. The conclusive remarks will follow after a brief chapter dealing with the drugs which have the potential to affect MCs in a tumour context.

# Chapter 2
# The Mast Cell

## 2.1 A Historical Overview

The controversial, and for long elusive, story of MCs begins on a remote day in the summer 1878, precisely on the 17th June of that year, when a 24-year-old medical student, the future Nobel Prize laureate Paul Ehrlich, discussed his doctoral thesis at the Medical Faculty of Leipzig University (Fig. 2.1). The title of his dissertation was: “Beiträge zur Theorie und Praxis der histologischen Färbung” (“Contribution to the theory and practice of histological dyes”) (Ehrlich 1878). In his presentation, dedicated to the chemical basis of the basic aniline dyes, the young Ehrlich devoted a chapter to the tissue staining properties of the aniline compounds and described for the first time a class of aniline-positive cells of the connective tissues endowed with cytoplasmic metachromatic granules (“granulierte Bindgewebezellen”) for which the name “Mastzellen” was proposed (Ehrlich 1878). This name, which means “well-fed cells”, was attached to the newly described cell population in the belief that their aniline-positive metachromatic granules might contain deposits of nutrients and might develop as a result of hypernutrition. He was likely to be mistaken about this issue, although it should be noted that recent observations in MC-deficient mice suggest roles for MCs in control of diet-induced obesity (Liu et al. 2009). Remarkably, Ehrlich stressed that the metachromatic staining of “Mastzellen” was due to reactivity of aniline with a “still undetermined chemical substance” stored in the granules (for further historical data, see Crivellato et al. 2003a). In addition, although he recognized that the “Mastzellen” were localized with very high frequency around blood vessels in the loose connective tissues, still these cells in his opinion had not to be regarded as members of a perivascular system. Their functional role, indeed, was mainly related to a “feeding” or “nourishing” activity, hence the term “Mastzellen”. Ehrlich found that aniline-reactive cells had a tendency to collect around developing preformed structures in connective tissues. For instance, he recognized that in certain acinar glands of the goat parotid, the pattern of “Mastzell” accumulation inside the tissue was not determined by the branching of the vascular system but by the ramification of the gland excretory ducts. He argued that “Mastzellen” could also be found around areas of developing

D. Ribatti, E. Crivellato, *Mast Cells and Tumours*,
DOI 10.1007/978-94-007-1469-4_2, © Springer Science+Business Media B.V. 2011

**Fig. 2.1** A portrait of Paul Ehrlich

tissues. As to the origin of these cells, he suggested that they might differentiate from fibroblasts.

In successive memories, Ehrlich studied the presence and significance of "Mastzellen" in pathological conditions. He described two situations where connective tissue might be over-nourished, in chronic inflammation, especially when this was aggravated by chronic lymphatic obstruction, and the environs of tumours. Here there existed a lymph stasis, a damming up of tissue fluid rich in nutriment, whereby certain fixed connective tissue cells were stimulated to become mobile, to multiply and to convert some of the abundant extracellular material into specific intracellular granules. According to Ehrlich, the "Mastzellen" were "indices of the nutritional state of the connective tissue" (Ehrlich 1879). They increased during periods of hypernutrition and diminished during periods of relative starvation. Interestingly, Ehrlich found many "Mastzellen" in tumours, especially carcinoma, but it was left to his pupil Westphal (1891) to recognize that the cells tended to accumulate at the periphery of carcinomatous nodules rather than within the substance of the tumour.

Ehrlich studied the staining reactions of blood cells, laying the foundations of modern haematology on the basis of the specific affinities of the leukocytes for various dyes (Ehrlich 1891; Ehrlich and Lazarus 1898). He encountered cells with basophilic, metachromatic granules, and thus came to recognize two types of "Mastzellen". The first—derived from, and living in the connective tissues (tissue "Mastzell"), the second—the counterpart of the neutrophil polymorph and eosinophil leukocyte—whose origin was in the bone marrow and whose habitat was in the peripheral blood (blood "Mastzell", basophil or mast leukocyte). Meanwhile Ehrlich (1891) had discovered basophilic granular cells in human blood, though so far only in myeloid leukaemia. Nevertheless, with characteristic insight he at once perceived that the blood MCs in higher vertebrates were true leukocytes stemming from precursors in the bone marrow. By the time that his textbook of 1898 came to be revised (Ehrlich and Lazarus 1909) the evidence for the myeloid origin of the blood MC was complete (Jolly 1900).

Sixteen years after Ehrlich's first description of Mastzellen, the English histologist and physiologist William Bate Hardy, provided a further contribution to the histochemical and functional definition of MCs. In the beginning, he referred to the formerly described Ehrlich's Mastzell with the collective term of "coarsely granular basophile cell" but in two outstanding papers, published in the Journal of Physiology in the years 1894 and 1895 (Kanthack and Hardy 1894; Hardy and Wesbrook 1895), he distinguished two types of granular basophile cells, i.e., the "coarsely granular basophile cells" and the "splanchnic basophile cells", which both belonged to the population of "wandering cells" (the modern leukocytes). These tissue-homing cells corresponded to the subsets of connective tissue-type and mucosal MCs, respectively, which would be described seventy years later by Enerbäck in rodents (Enerbäck 1966a, b). Among the coarsely granular basophile cells, he also differentiated those cells which populated the serosal cavities—the so-called coelomic coarsely granular basophile cells—from the common coarsely granular basophile cells which were localized in the connective tissues. He stated that the granular basophile cells presented with different morphological and histochemical characteristics in diverse animal species as well as at different anatomical sites, being thus the first scholar to shape the fundamental concept of "MC heterogeneity". Hardy performed a series of functional experiments on the basophile cells and suggested that different granular basophile cells might express functional specialisations. Hardy's view of basophile cell function was partly in line with Ehrlich's concept of a nutritional role for these cells. He believed that these cells might be somehow involved in the up-take and storage of substances as a result of hypernutrition. However, he also explored other experimental areas, such as the potential contribution of granular basophile cells to phagocytosis of pathogens and the participation of these cells to defence mechanisms during infections (for further historical data, see Crivellato and Ribatti 2010a).

For the first 60 years after Ehrlich's discovery of MCs, the study of these cells was almost entirely histological. MCs were demonstrated in most animal species, although with an irregular, capricious distribution (Michels 1938). In his for long unsurpassed review of the MCs, Michels praised Ehrlich's pioneering contribution to the study of these cells, in particular his recognition that MC granules were soluble in water. He wrote: "uncounted pages of useless and misleading research have been the result of the failure on the part of many investigators to heed the admonition originally given by Ehrlich and Westphal, that the MC granules are soluble in water and that to preserve them tissues must be fixed in 50% alcohol and stained in alcoholic thionine" (Michels 1938). As for the relation of MCs with basophils, after an initial unitary conception, subsequent studies indicated that at least in higher organisms these two cells differed both in habitat and in parentage, being the derivation of MCs unknown—they were usually interpreted as istiogenic elements—whilst the origin of basophils was from the bone marrow. Michels wrote that "aside from an identical basophilic metachromatic reaction of the granules, the two cell types have nothing in common" (Michels 1938).

At the end of the 1930s, a group of Scandinavian researchers provided fundamental new insight as to MC structural and functional profiles. The mysterious MC component prophesized by Ehrlich as the responsible agent for granule metachromasia was revealed by Jorpes, Holmgren and Wilander (Holmgren and Wilander 1937; Jorpes et al. 1937). Following Jorpes' discovery that the anticoagulant heparin—a polysulphuric acid ester, made up of glucuronic acid, glucosamine and sulphuric acid—was subject to stain metachromatically with toluidine blue, Holmgren and Wilander reconsidered Ehrlich's observation that MC granules stained metachromatically with toluidine blue. These authors were able to set a correlation between the number of toluidine blue-positive MCs in various tissues and their heparin content. Tissues with large amounts of "Ehrlichschen Mastzellen" were particularly rich in heparin and, among MC-rich tissues, the beef liver capsule was described as "a pure culture of MCs" (Holmgren and Wilander 1937; Jorpes et al. 1937). The Swedish investigators formulated the conclusive theory that the task of MCs in the connective tissues was to produce heparin.

The discovery that tissue MCs were the source of heparin was the prelude to the identification of two other crucial substances contained in MCs: histamine and serotonin. A potential correlation between tissue heparin and tissue histamine contents was initially established. Indeed, the release of histamine was shown to be accompanied by a similar release of large amounts of heparin both *in vivo* and *in vitro* (Rocha and Silva 1952). In a series of fundamental studies published in the period 1952–1956, the pharmacologists Riley and West in Scotland demonstrated that histamine—the previously identified Lewis' "H substance" responsible for skin anaphylactic phenomena—was present in MCs. Early studies showed that injection in the rat of histamine liberators, such as stilbamidine and D-tubocumarine, was followed by selective damage to tissue MCs indicating that MCs were the presumptive site of histamine accumulation in the tissues (Riley and West 1952). Further investigations revealed that very high values for heparin and histamine could be found in tissues which were exceptionally rich in MCs, such as the cleaned capsule of normal ox liver, and the sheep and ox pleura (Riley 1955). The loose connective tissues but not the dense connective tissue of the tendons were rich in histamine and MCs as well. A strong positive correlation between histamine tissue contents and the histological demonstration of MCs was also recognized in pathological conditions such as urticaria pigmentosa in man (Riley and West 1953) and MC tumours in dogs (Cass et al. 1954). Further evidence for the presence of histamine in MCs was provided by Fawcett and by Mota and Vugman. Fawcett (1954) demonstrated that the potent histamine liberator, compound 48/80, caused release of MC granules, and that it failed to liberate appreciable amounts of histamine from connective tissue previously depleted of MCs. Mota and Vugman (1956) reported a good correlation between serum histamine levels and disruption of MCs in a guinea pig model of anaphylaxis. On the whole, these data show that by the end of the 1950s experimental studies had delineated a fundamental functional link between MCs, histamine and the allergic and anaphylactic reactions which had been recognized and described long before by Pirquet, who coined the term "allergy" in 1906, and Portier and Richet, who introduced the term "anaphylaxis" in 1902. Thus, MCs could since be defined as the major tissue repository for histamine ("histaminocytes") and were entitled to play a crucial role in allergic conditions such as hay fever, asthma and anaphylactic shock (for further historical data, see Beaven 2009). It soon appeared, however, that basophils too were rich in histamine and heparin (Graham et al. 1955; Behrens and

Taubert 1952). Thus, the similarities between MCs and basophils seemed again to outweigh their differences (Riley 1954).

About the same period, Benditt and co-workers demonstrated that 5-hydroxytryptamine (5-HT, serotonin) associated with MCs of the subcutaneous areolar tissue of the rat (Benditt et al. 1955). Serotonin was a vasoconstrictor substance suspected for decades to be contained in platelets. It was isolated and characterized in 1948 by Maurice Rapport and Irvine Page, and discovered to correspond to enteramine by the Italian scientist Erspamer in 1952 (Rapport et al. 1948; Erspamer and Asero 1952). Later work by Parratt and West (1956) revealed that serotonin was concentrated in tissue MCs of the rat and the mouse but not of the guinea-pig, dog, man, rabbit, cow, hamster and cat, and that the skin of the rat contained more than half of the total serotonin of the body. These authors speculated that MC serotonin might be involved in the response of animals to injections of large-molecular-weight substances such as egg-white or dextran. Both histamine and serotonin had potent effects, especially on the vascular system, and their release by MCs, which expressed a preferential perivascular location, could be implicated in the inflammatory reactions occurring in connective tissues (Riley 1963). Remarkably, not only MCs had the ability to synthesize histamine and serotonin but, as demonstrated by Green's group, both normal and neoplastic MCs were able to take-up exogenous histamine and serotonin (Day and Green 1962; Furano and Green 1964). It was thus made clear that MCs can concentrate biogenic amines.

Hardy's concept of MC heterogeneity was further developed in the 1960s by Enerbäck. Based on their specific staining characteristics and preferential tissue homing, two morphologically distinct subpopulations of rodent MCs were initially identified and termed connective tissue MCs (CTMCs) and mucosal MCs (MMCs), respectively (Enerbäck 1966a, b, 1986). The former populated the mucosae of the respiratory and gastrointestinal tracts, while the latter homed to the connective tissues and serosae. CTMCs could be distinguished from MMCs by staining in red with safranin due to the presence of large amounts of heparin in their secretory granules. In the mouse, indeed, the proteoglycan content of MC granules varied in the different MC subtypes. CTMCs contained heparin that lacked in MMCs. Conversely, MMCs expressed chondroitin sulphates A and B, which were not found in CTMCs, whereas both MC subtypes stored chondroitin sulphate E in their granules. Thus, in contrast to CTMCs, MMCs were sensitive to routine formalin fixation and could not be identified in standard histological sections. After appropriate fixation and sequential staining with Alcian blue and safranin, the MMCs stained blue, being thus differentiated from CTMCs which stained with safranin and were red. It later appeared that CTMC and MMC subtypes contained distinct classes of proteases and expressed different functional profiles, being activated in part by different stimulators and providing selective secretory responses.

During the 1970s, other crucial aspects of MC involvement in allergic and anaphylactic reactions were recognized. It was not until 1967 that the "reaginic" antibody—the transferable factor responsible for the sensitization phenomenon described by Prausnitz in 1921 (Prausnitz and Küstner 1921)—was eventually identified by Kimishige Ishizaka and Teruko Ishizaka as γE-antibodies (immunoglobulin

E, IgE), a minor component of the Ig family (Ishizaka and Ishizaka 1967). IgE was shown to be capable to mediate the release of histamine and another mysterious substance called "slow reacting substance of anaphylaxis" (SRS-A) from sensitized tissue MCs (Ishizaka et al. 1970; Orange et al. 1971). The receptor for IgE molecules was later identified at high concentrations on the surface of MCs and was recognized to bind IgE with high affinity and specificity. This was named the "high affinity" receptor for IgE (FcεRI), and it was completely cloned in 1989 (Blank et al. 1989). The cross-linking of IgE with bivalent or multivalent antigen on the surface of MCs resulted in the aggregation of IgE receptors and in the triggering of MC degranulation. SRS-A was initially recognized by Feldberg and Kellaway in 1938 as a spasmogenic substance distinct from histamine (Feldberg and Kellaway 1938). This stuff was capable to increase microvascular permeability and produce long-lasting wheal-and-flare responses in the skin and bronchoconstriction in the lungs during anaphylactic shock (Brocklehurst 1960). It became apparent that SRS-A was a mixture of lipids, collectively named leukotrienes (LTs), which were generated from arachidonic acid through the 5′-lipoxygenase pathway. MCs were recognized to synthesize $LTC_4$, $LTD_4$ and $LTB_4$ as well as other inflammatory lipids (eicosanoids) called prostaglandins (PGDs) (Roberts et al. 1979). Within minutes of MC stimulation with anti-IgE antibodies, which activate the IgE receptor on the cell surface and trigger MC response, MCs were seen to release substantial amounts of $PGD_2$ (Lewis et al. 1982). Parallel investigations showed that basophils were also endowed with the high-affinity receptor for IgE, a discovery that further linked the two cell lineages in the speculative approach of researchers. In addition, both MCs and eosinophils were recognized to participate to certain protective reactions against parasites (Askenase 1977). Thus, until the mid 1990s, the paradigm prevailed that MCs were to be regarded as tissue elements principally implicated in the pathogenesis of allergic reactions and responsible for defence against certain parasites.

In the 1970s, accurate definition by transmission electron microscopy of a series of MC structural and functional details was provided by several groups. In particular, the group of Ann Dvorak in Boston afforded persuasive demonstration of the heterogeneous structure of MC granules, clarifying the mechanisms of MC degranulation and recovery, and providing new data on the fine, distinctive aspects of MC and basophil ultrastructure. During MC degranulation, cytoplasmic granule membranes were seen to fuse with each other and with the plasma membrane, giving rise to open secretory channels which allowed the release of granule contents into the extracellular environment (Dvorak 1991). This quick, explosive, IgE-mediated process of MC degranulation, characteristic of type I hypersensitivity reactions and categorized as "anaphylactic degranulation" or "compound exocytosis", was punctually differentiated by a novel pattern of MC secretion Dvorak's group was able to identify, for which the term "piecemeal degranulation" was coined. In a series of elegant ultrastructural immunocytochemical experiments, Ann Dvorak and co-workers described the fine aspects of this newly-identified degranulation pathway, which allowed MCs to release subtle amounts of granule-stored material in a prolonged time lapse. There was a slow discharge of granule contents in a "piecemeal" fashion, without membrane fusion events and granule opening to the cell exterior. Dvorak underlined the concept that "piecemeal degranulation" represented the most common way of MC secretion observed in MCs infiltrating areas of chronic inflammation or tumours whilst the well studied and much more renowned pattern of "anaphylactic degranulation" could rarely be recognized apart from the sites of allergic responses (Dvorak 1991).

The primary involvement of MCs in such so harmful and sometimes life-threatening events as allergic and anaphylactic reactions left researchers somewhat disconcerted as to the real physiological role of MCs. How was it possible that so a ubiquitous and universally distributed cell type might exist only to cause danger to the host? Still in 1975, Harold Dvorak and Ann Dvorak wrote: "much more must be learned before we can confidently describe the role of basophils, or of the closely related MCs, in health or disease. It seems most unlikely that either cell exists for the purpose of destroying the organism in anaphylactic shock. Nonetheless, it is highly probably that basophil/MC function is closely related to the potent chemicals stored within their cytoplasmic granules. One likely possibility holds that small amounts of these chemicals are required for homeostasis (e.g., for regulation of the tone of the microvasculature) and that these cells function by releasing such substances continuously, as they are needed, in small aliquots rather than by explosive discharge" (Dvorak and Dvorak 1975).

By the end of the 1970s, scientists were able to solve the long-lasting enigma of the origin of MCs. Demonstration of MC derivation from bone marrow precursors could be established in 1977 when Kitamura's group first showed that, using the abnormal giant cellular granules of beige mice (C57BL-$Bg^J/Bg^J$) as a traceable marker (Fig. 2.2), tissue MCs were found to develop from grafted beige bone marrow in irradiated wild type recipient mice (Kitamura et al. 1977). This discovery prompted further investigations on the origin of MC lineage leading to the concept that MCs were tissue-homing leukocytes. The paradigm was developed that they arise from pluripotent haematopoietic stem cells, circulate in the blood as agranular progeni-

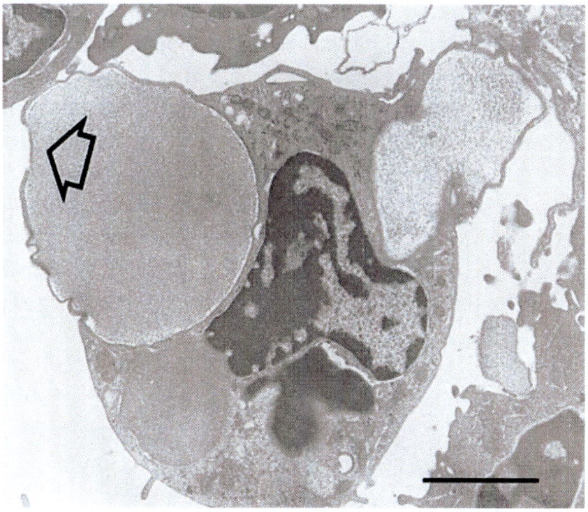

**Fig. 2.2** Transmission electron microscopy of a mast cell with abnormal giant granules from a beige mouse (C57BL-BgJ/BgJ). The mast cell has been stimulated to exocyte and the granule on the *left* (*arrow*) is opening to the cell exterior. Bar: 3 μm. (Reproduced from Crivellato et al. 1999)

tors, and then acquire their mature phenotype within tissues. The main growth factor governing tissue MC development and complete differentiation was shown to be the KIT receptor ligand or stem cell factor (SCF). The importance of SCF as a MC growth factor was underlined by the fact that mice with certain loss-of-function mutations affecting either SCF or its receptor KIT were devoid of MCs. Indeed, lack of expression of a functional KIT receptor due to spontaneous mutation in both copies of *Kit*, as it occurred in genetically MC-deficient WBB6F1-$Kit^{W}$-$Kit^{W-v}$ mice ($W/W^{v}$ mice), resulted in a virtual absence of tissue MCs (Kitamura et al. 1978). This important finding stimulated further studies on the genetics of the KIT-SCF system. These studies provided a series of MC-lacking mouse strains which revealed extremely useful to study different aspects of MC function. Indeed, lack of MCs in *Kit*-mutant mice could be selectively repaired by the adoptive transfer of genetically compatible wild-type or mutant MCs derived from *in vitro* cultures to create the so-called MC "knock-in" mice (Nakano et al. 1985). Most of our current knowledge on MCs is actually derived from MC "knock-in" mouse models which have allowed researchers for testing and verifying whether MCs contribute to specific functions. By the way, the finding that basophils lacked the KIT receptor and were unaffected by the SCF shaped the concept that the developmental pathways of MCs and basophils were different.

In the 1980s, many investigations focussed on the composition of MC granules and the capacity of stimulated MCs to release cytokines, chemokines and growth factors. The concept of MC heterogeneity was further defined at this stage of MC research (Bienenstock et al. 1983). Like rodent MCs, human MCs were also found to express morphological, biochemical and functional heterogeneity. The first evidence that MCs contained proteases was provided by Gomori in 1953 (Gomori 1953). He developed enzyme histochemical techniques for detecting esterase activity inside of cells in sections of fixed tissues and was able to recognize that MCs stained intensely with such procedures. By 1960, two proteases with chymotrypsin- and trypsin-like activity were identified in MCs (Benditt 1956; Benditt and Arase 1959; Glenner and Cohen 1960). Enzyme activity was recognized to localize within intracellular granules. These enzymes were purified in the 1980s and renamed tryptase and chymase (Schwartz et al. 1981; Schechter et al. 1986). It soon appeared that MCs from different anatomical sites contained different profiles of these enzymes as well as of other proteases identified in the meantime. Human MCs were thus divided into two subtypes depending on the expression of different proteases in their granules (and other functional features) (Irani et al. 1986). MCs, which contained tryptase only, were designated as $MCs_{T}$ or "immune cell-associated" MCs. They were predominantly located in the respiratory and intestinal mucosa, where they co-localized around T lymphocytes. MCs that contained both tryptase and chymase, along with other proteases such as carboxypeptidase A and cathepsin G, were referred to as $MCs_{TC}$. They were predominantly found in connective tissue areas, such as skin, submucosa of stomach and intestine, breast parenchyma, myocardium, lymph nodes, conjunctiva and synovium. These two subsets of human MCs differed also in terms of their mediator content and reactivity. A third type of MC, called $MC_{C}$ was also identified. This MC expressed chymase without tryptase and resided mainly in the submucosa and mucosa of the stomach, small intestinal submucosa and colonic mucosa (Irani and Schwartz 1994). Interestingly, human $MCs_T$ were seen to correspond most closely to rodent MMCs, whereas $MCs_{TC}$ resembled rodent CTMCs. It was later recognized that the concept of MC heterogeneity was not limited to staining properties but also involved functional characteristics.

Beginning from the end of the 1980s, it progressively emerged that granules in MCs contained a series of highly active biological compounds, such as cytokines, chemokines and growth factors. In 1989, a series of groups investigating on MC responses to various activators reported that stimulated MCs produced and released interleukin (IL)-3, IL-4, IL-5, IL-6 and granulocyte/macrophage-colony stimulating factor (GM-CSF) (Burd et al. 1989; Plaut et al. 1989; Wodnar-Filipowicz et al. 1989) and that this could occur in the absence of MC degranulation. Shortly thereafter, Gordon and Galli (1990) reported that MCs were a biologically relevant source of both preformed and antigen-induced tumour necrosis factor (TNF)-α. These discoveries were central to set MCs in the midpoint of a complex series of inflammatory and immunological pathways associated to host defensive responses, as revealed by two seminal papers appearing in 1996 which demonstrated that MCs were essential to survival in a mouse model of sepsis (Malaviya et al. 1996; Echternacher et al. 1996). On the basis of these and other succeeding findings, it appeared that MCs were endowed with a large series of preformed and newly-synthesized mediators capable to exert different immunological and non-immunological functions. But this is a much recent story that will be accounted for in the next chapters of this book.

## 2.2 Biology of Mast Cells

### *2.2.1 Mast Cell Structure and Ultrastructure*

MCs are bone marrow-derived tissue-homing secretory cells, which have been identified in all vertebrate classes. A cell population with the overall characteristics of higher vertebrate MCs is identifiable even in the most evolutionarily advanced fish species (Crivellato and Ribatti 2010b). Overall, MCs present consistent species-specific differences as to shape, dimension and granule staining properties. Besides, MCs coming from different sites in the same species or even the same organism still reveal subtle structural and ultrastructural specializations. We will here briefly sketch the microscopical traits of human and rodent MCs. Viewed by light microscopy, human MCs usually present as round or elongated cells with a diameter ranging between 8 and 20 μm, depending on the organ examined. Their single nucleus shows a round or oval shape and the cytoplasm contains numerous secretory granules that metachromatically stain with thiazine dyes such as toluidine blue (Fig. 2.3). By electron microscopy, these cells exhibit a non-segmented monolobed nucleus with peripherally condensed chromatin. The cytoplasm contains a few mitochondria, short profiles of the rough endoplasmic reticulum and numer-

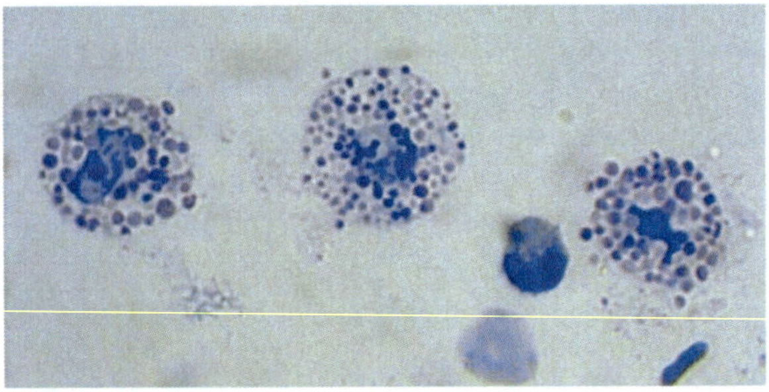

**Fig. 2.3** Semi-thin section showing three mast cells isolated from the rat peritoneal cavity and stained with toluidine *blue*. Numerous cytoplasmic metachromatic granules are recognizable. Original magnification × 1,000. (Reproduced from Ribatti et al. 2001b)

ous free ribosomes. The most characteristic cytoplasmic organelles in human MCs are the membrane-bound, moderately electron-dense secretory granules. Secretory granules are very abundant and correspond to the metachromatic granules seen at the light microscopy level. They have an average diameter of 1.5 μm and present different types of substructural patterns, i.e., homogeneous, crystalline, scroll, particle, thread-like or a combination of them (reviewed by Dvorak 1991) (Figs. 2.4 and 2.5). Granule ultrastructure has been partly related to its content of serine proteases.

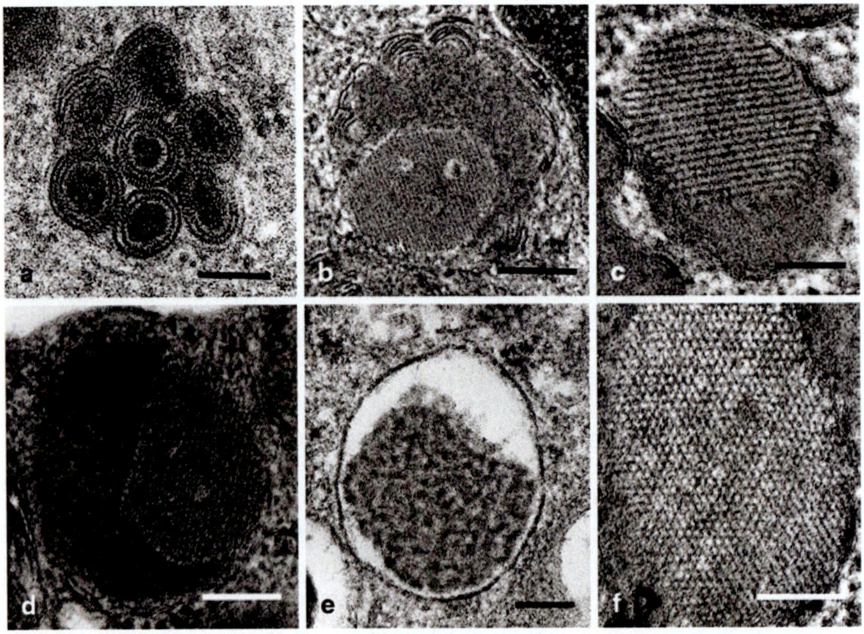


**Fig. 2.4** Ultrastructural morphology of human mast cell granules. Different patterns of granule texture may be recognized by transmission electron microscopy. In (**a**) a typical scroll granule from a duodenal mucosa mast cell. In (**b**) a mixed scroll-homogeneous-crystal granule from a mast cell in the skin. (**c**) and (**f**) show two crystal granules with very regular parallel arrays and hexagonal arrays, respectively, from skin mast cells. In (**d**) a mixed dense-crystal granule from a lung mast cell. In (**e**) a particle granule from a lymph nodal mast cell. Bars: 0.1 μm

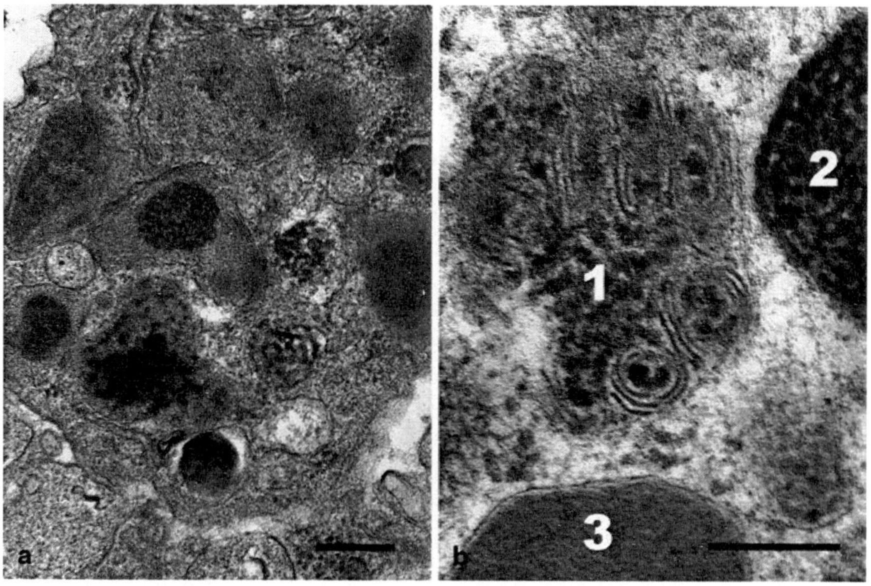


**Fig. 2.5** Examples of mixed granules in a lymph nodal mast cell. In (**a**) mixed granules mainly contain coarse particles and scrolls. Viewed at higher magnification (**b**) three types of granules are depicted, a mixed scroll-particle (1), a particle (2) and a homogeneously dense (3) granule. Bars: 0.3 μm in (**a**), 0.1 μm in (**b**). (Reproduced from Crivellato et al. 2003c)

Indeed, granules with the chymase protease preferentially exhibit homogeneous or crystalline substructures whereas granules lacking this protease show mainly a scroll pattern. However, significant granule heterogeneity can be found in any particular tissue and even between granules of a single MC. Aside from the typical secretory granules, human MC also contain non-membrane-bound, highly osmiophilic granules, called lipid bodies (Dvorak 1991). Lipid bodies are prominent in human MCs but less frequent in mouse, rat and guinea pig MCs. They are fewer in number and generally larger than secretory granules, and serve as a significant storage site for arachidonic acid. Recently, both secretory granules and lipid bodies in human MCs have been implicated in RNA metabolism (Dvorak and Morgan 2000; Dvorak et al. 2003).

Rodent MCs may consistently differ from human MCs. In guinea pig MCs, the substructural complexity of granule patterns resembles that of human MCs. Viewed by transmission electron miscoscopy, granules present either regular crystalline arrays, or arrays of tubules, or irregular thick threads, or finely granular material, or mixtures of these patterns (Dvorak 1991). Conversely, the granules of mouse MCs are filled with homogeneous dense material. Their content does not show the considerable variations in substructure that are seen in both human and guinea pig MCs.

Being MCs highly differentiated cells endowed with plentiful of such secretory organelles, research has much focussed on the mode of granule discharge. Viewed by electron miscoscopy, MCs present two morphologically distinct patterns of content release from granule stores, called: (1) compound exocytosis, also referred to as anaphylactic degranulation, and (2) piecemeal degranulation.

Anaphylactic degranulation was the first strudied mechanism of MC secretion. It is characteristically triggered by aggregation of FcεRI due to cross-linking of IgE with bivalent or multivalent antigens. This specific stimulus elicits the explosive release of granules and the liberation of preformed mediators contained inside. The term of anaphylactic degranulation has been attached to this kind of secretory sequence because it is characteristic of the rapid discharge of MC products occurring during anaphylaxis. The distinct phases of this process have been carefully studied by transmission electron microscopy in various experimental conditions (reviewed by Dvorak 1991) (Fig. 2.6). Activated granules initially swell, exhibit matrix dissolution and decreased density. The perigranule membranes of adjacent granules fuse with each other and with the plasma membrane, resulting in the formation of elongated, tortuous chains of interconnected granules, called secretory channels, which eventually open at the surface. These openings allow rapid release of histamine and other low-molecular-weight substances stored in the secretory granules. All these events are very rapid, being ultrastructurally demonstrable as early as 3 minutes after MC stimulation. Degranulating MCs show profound alterations of their plasma membrane profile, being covered with complex elongated folds and surface projections. Later, they exhibit even shedding of cellular membranes. It soon became ap-

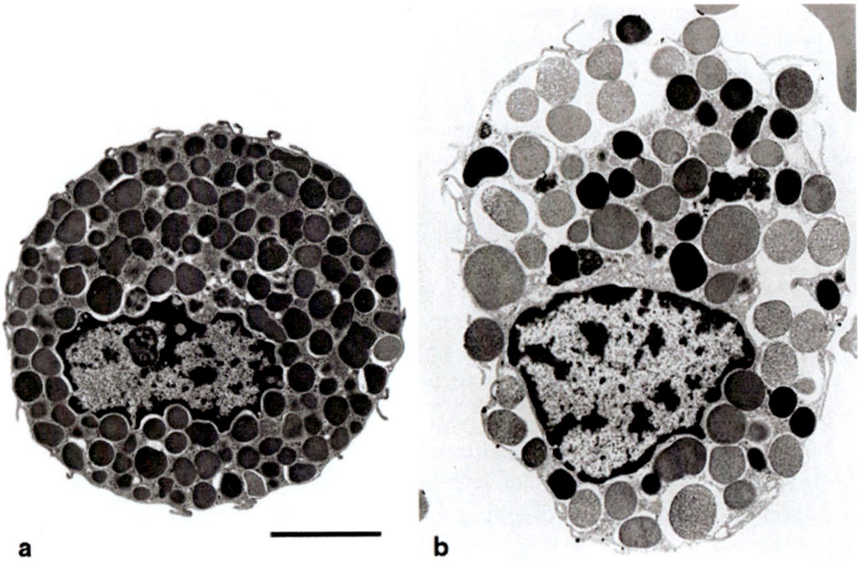


**Fig. 2.6** Transmission electron microscopy of rat peritoneal mast cells. In (**a**) the mast cell shows the ultrastructural morphology of a resting cell, with smooth contour and a full complement of cytoplasmic secretory granules. In (**b**) the mast cell is in a state of active exocytosis and shows elongated surface folds, extruded, membrane-free granules, and cytoplasmic secretory channels filled by swollen granules. Bar equals 3 μm in (**a**) and 1 μm in (**b**). (Reproduced from Crivellato et al. 1999, 2002b)

parent, however, that although absolutely typical in its ultrastructural characteristics and highly significant in its dramatic functional consequences, anaphylactic degranulation of human MCs was rarely encountered in *in vivo* conditions. Partial forms of exocytosis, involving release of a limited number of granules, can also be seen.

Besides anaphylactic degranulation, a different type of cell secretion termed piecemeal degranulation was identified in MCs by Ann Dvorak and co-workers in the early 1970s of the last century. This novel form of granule discharge was first recognized in guinea pig and human basophils participating to skin contact allergic reactions or infiltrating tumours but it was soon detected in MCs as well (Dvorak et al. 1973, 1974; Dvorak 2005a, b). When examined by transmission electron microscopy, cells exhibited partially or completely emptied granules in the absence of granule-to-granule or granule-to-plasma membrane fusions (Fig. 2.7). It appeared that piecemeal degranulation was substantially different from anaphylactic degranulation, which effected extrusion of stored material through formation of secretory channels and granule fusion to the cell plasma membrane. The term piecemeal degranulation was coined because cytoplasmic granules showed focal pieces or packets of lost particles leaving characteristic patchy areas of electron-density beside

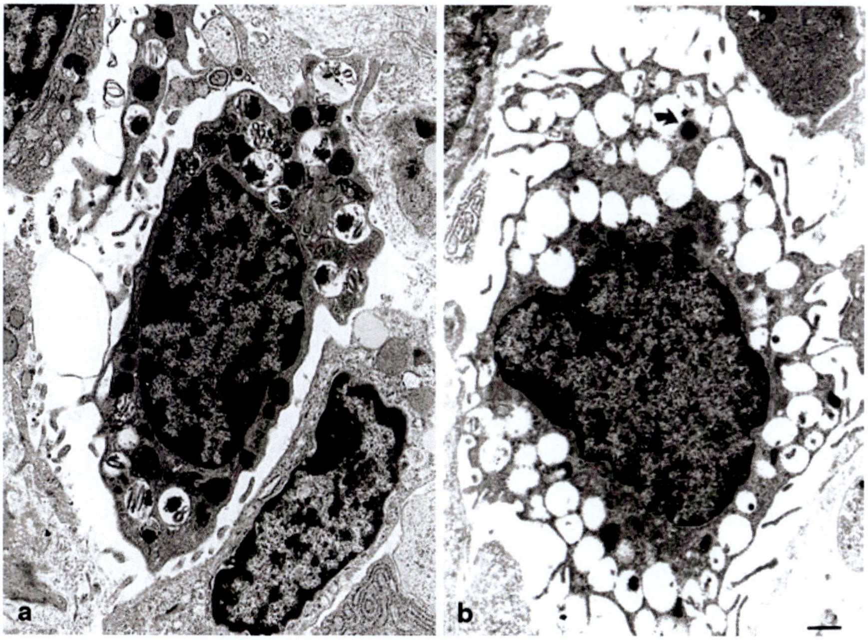


**Fig. 2.7** Transmission electron microscopy of two mast cells exhibiting different stages of piecemeal degranulation. The mast cell in (**a**), taken from the human intestinal mucosa, is in an active secretory condition and shows a number of expanded, non-fused secretory granules which exhibit characteristic loss of matrix structure. In the mast cell in (**b**), which is taken from the human skin, the piecemeal degranulation process is in a virtual final stage and the cell cytoplasm is filled by non-fused, almost empty secretory containers. The arrow points to an osmiophilic granule (lipid body). Bars: 1 µm in (**a**), 1.5 µm in (**b**). (Reproduced from Crivellato et al. 2003b, 2004)

luciency zones. In addition to granule changes, the cytoplasm of activated MCs presented a large number of smooth, 30–150 nm in diameter, membrane-bound vesicles. Some of these vesicles were filled with particles similar in structure and electron-density to those contained in the granules while others were apparently empty and appeared electron-lucent. Remarkably, vesicles were often seen attached to granules, in a process of budding from or fusing with the perigranule membrane. The piecemeal degranulation phenotype was recognized in MCs localized at sites of chronic inflammatory responses, and a careful scrutiny of MC morphology in different human pathologies led to the conclusion that piecemeal degranulation, not anaphylactic degranulation, was the most common release reaction identified in these cells (reviewed in Dvorak 1991). Dvorak and co-workers provided a theoretical model to explain how granules would empty during this kind of releasing reactions and gave rigorous kinetic bases to the movement of granule content within individual cells. The "shuttling vesicle" hypothesis formulated by Dvorak and Dvorak (1975) postulated a vesicular transport mechanism to effect transfer of granule constituents outside the cells. This hypothesis received substantial experimental confirmations by a series of elegant electron microscopic investigations using ultrastructural tracers and purified cells stimulated *in vitro* by different secretagogues. According to this model, an outward flow of cytoplasmic vesicles loaded with granule materials effects granule emptying during piecemeal degranulation. Vesicles containing bits of granule contents bud from the perigranule membrane, move through the cytoplasm and fuse with the plasma membrane, leading to content discharge. Endocytic vesicles are retrieved from the plasma membrane, traverse the cytoplasm and fuse with granules in a closely coupled inward flow. If the rate and amount of vesicular traffic are balanced, granule containers empty in a piecemeal fashion but maintain a constant size. If, on the other hand, the inward flow of the endocytic vesicles exceeds the outward flow of the exocytic vesicles, the granule chambers become enlarged. The latter event is what generally occurs during piecemeal degranulation. The electron microscopic changes observable during piecemeal degranulation can be summarized as follows: (1) Being piecemeal degranulation a discrete process affecting single granules in an asynchronic, stepwise progression, what generally results is a unique granule polymorphism, which consists of an admixture of normal resting granules, activated granules with enlarged chambers and diminished constituents, and empty dilated containers. (2) Remarkably, each granule does not fuse with the others or with the cell membrane but maintains its close individual structure during the entire releasing process. (3) The residual secretory material contained in activated granules presents "piecemeal" loss of constituents leading to "semilunar" or "haloed" patterns. (4) A proportion of granules exhibits surface budding projections; these are either apparently empty (thus they appear electron-lucent) or filled by the same electron-dense material that constitutes the granule. (5) Small, smooth, membrane-bound, electron-dense or -lucent vesicles are recognizable attached to the granules or free in the intergranular cytosol or close to the plasma membrane. Thus, identification of piecemeal degranulation relies upon specific ultrastructural criteria, which refer to both granule and cytoplasmic changes.

Functionally speaking, anaphylactic degranulation and piecemeal degranulation are opposite events. On the one hand, anaphylactic degranulation is a rapid and massive process of cell secretion which allows for complete discharge of granule constituents, often even of granule matrices. It does not accommodate the intensity of the releasing process and does not enable the cell to effect a differential discharge of granule-stored products. On the other hand, piecemeal degranulation is a slow, long lasting event, which is likely to permit the cell to single out a definite substance or a limited number of substances from the miscellaneous pool of releasable constituents packed inside secretory organelles. Evidence has been accumulated indicating that piecemeal degranulation in MCs allows for differential release of granule constituents. In 1990, Askenase's group published an important article that provided indication for a differential release of serotonin without histamine and without anaphylactic degranulation from rat peritoneal MCs pretreated with the tricyclic antidepressant drug amitriptyline and stimulated with compound 48/80 (Kreuter Kops et al. 1990). Viewed at the electron microscope, storage granules lost their homogeneity, exhibited greatly reorganized matrix and were surrounded by clear spaces which were often associated with small (10–100 nm diameter) cytoplasmic vesicles, some of which contained electron-dense material. Secretory granules often had bud-like protrusions. The general pattern of MC secretion corresponded to what had already been known with the term of piecemeal degranulation. As subsequently demonstrated by different groups, piecemeal degranulation can be conceptualized as a tuneable process, which may be triggered and modulated by distinct compounds and which may account for the "surgical" cytokine response exhibited by MCs in different functional settings (Theoharides et al. 2007).

## *2.2.2 Origin, Development and Tissue Homing of Mast Cells*

The MC is a type haematopoietic cell which acquires its definite phenotype once entered the homing tissues. These cells originate from progenitor cells in the bone marrow, which move through the circulation and become mature MCs after homing to destination tissues under the influence of the local microenvironment (Gurish and Austen 2001; Kitamura and Ito 2005).

Kitamura et al. (1977) first showed in the mouse that MCs derive from bone marrow precursors. Using the abnormal giant cellular granules of beige mice (C57BL-$Bg^J/Bg^J$) as a traceable marker, they found that tissue MCs developed from grafted beige bone marrow in irradiated wild type recipient mice. In humans, MCs derive from $CD34^+$, $CD13^+$, $Fc\varepsilon RI^-$, $KIT^+$ committed progenitors (Kirschenbaum et al. 1991). Committed progenitors, circulating as agranular mononuclear leukocytes, traverse the vascular space and complete their maturation after moving into diverse peripheral tissues (Rodewald et al. 1996). Here, they acquire concomitant phenotypic diversity. Rodewald et al. (1996) first identified the committed progenitors for the MC lineage in mouse fetal blood. This MC progenitor was defined by the surface phenotype Thy-$1^{low}$ $KIT^{high}$, lacked the expression of FcεRIα transcript and contained cytoplasmic granules. MC colony-forming cells reside within the bone marrow, spleen, peripheral blood, mesenteric lymph nodes and gut mucosa (Crapper and Schrader 1983). MC progenitors, bearing the phenotype $Lin^-KIT^+Sca\text{-}1^-Ly6c^-Fc\varepsilon RI\alpha^-CD27^-\beta7^+T1/ST2^+$, were identified in the adult bone marrow (Chen et al. 2005). These cells develop into MCs in culture and reconstitute MC compartment upon their transplantation into MC-deficient mice. Recently, a cell population ($Lin^-Kit^+Fc\gamma RII/III^{hi}\beta7^{hi}$) has been identified in the mouse spleen with the characteristics of a bipotent progenitor for the basophil and MC lineages (Arinobu et al. 2005). This cell population, termed basophil/MC common progenitor, can be generated mainly from granulocyte/macrophage progenitors in the bone marrow. The same authors identified MC progenitors ($CD45^+Lin^-CD34^+\beta7^{hi}Fc\varepsilon RI\alpha^{lo}$) in the intestine.

The developmental pathway of MCs and the relationship between MCs and other leukocytes are controversial (Arinobu et al. 2009). It has been debated whether MCs are of "myeloid" or "lymphoid" origin or stem from a distinct population of precursor cells. Specific arrays of differentiation factors such as SCF and IL-3 expressed by bone marrow stromal cells promote a distinctive pattern of MC-specific gene expression that includes genes encoding for distinct transcription factors (Winandy and Brown 2007). The main regulatory pathway for normal MC differentiation is not well characterized. Balanced activity of transcription factors PU.1 and GATA is known to be required, along with the transcription factors Mitf and possibly SCL, and the functions of GATA-2 and GATA-1 in this process can now be distinguished (Arinobu et al. 2005; Babina et al. 2005; Nishiyama et al. 2005). Until recently, it has remained obscure whether MCs are "myeloid" or are in a distinct class of their own. They can be derived from precursors separate from most myeloid lineages (Arinobu et al. 2005; Chen et al. 2005). Unlike monocyte and granulocyte lineages, but corresponding to lymphocytes, they develop through a pathway that excludes transcription factor C/EBP-α, which controls monocyte/dendritic cell programs (Iwasaki et al. 2006; Taghon et al. 2007). Recently, results have provided concrete evidence for one specific function, increased GATA-2 or GATA-3 expression, that provides direct access to the MC pathway for an uncommitted but differentiating lymphoid precursor (Taghon et al. 2007).

The zinc finger transcription factor GATA-1 has crucial roles in erythroid, megakaryocytic, and MC differentiation. Remarkably, friend of GATA-1 (FOG-1) is a binding partner of GATA-1 and is indispensable for the function of GATA-1 during erythro/megakaryopoiesis, but FOG-1 is not expressed in MCs (Cantor et al. 2008). A combined experimental system with conditional gene expression and *in vitro* haematopoietic induction of mouse embryonic stem cells has shown that expression of FOG-1 during the progenitor period inhibits the differentiation of MCs and redirects them into the erythroid, megakaryocytic, and granulocytic lineages (Cantor et al. 2008; Sugiyama et al. 2008). Mutant analysis reveals that this lineage skewing is caused by disrupted binding between GATA-1 and PU.1, a transcription factor that positively or negatively cooperates with GATA-1, which is a prerequisite for MC differentiation (Sugiyama et al. 2008). However, FOG-1 expression in mature

MCs brings approximately a reversible loss of the MC phenotype. Thus, FOG-1 inhibits MC differentiation in a differentiation stage-dependent manner and its down-regulation is a prerequisite for MC development (Cantor et al. 2008; Sugiyama et al. 2008). Although the main function of GATA-3 is to act as a master transcription factor for the differentiation of T helper (Th) 2 cells, new research indicates that GATA-3 too is a crucial factor for MC development. It has been demonstrated that expression of GATA-3 in the absence of Notch-DL1 signalling drives MC development (Taghon et al. 2007). Indeed, overexpression of GATA-3 in thymic progenitor cells promotes MC differentiation. By the way, the thymus has been recognized to be the site of intense MC proliferation during chick embryo development (Crivellato et al. 2005a; Fig. 2.8).

Once dismissed from the bone marrow as well as other embryonic sites of haemopoiesis, MC precursors move through the blood circulation to peripheral target sites. Tissue homing of MC precursors is critically regulated by tissue micro-environmental factors. Among these, the SCF secreted by fibroblasts, stromal cells and endothelial cells represents the most important cytokine involved in human and rodent MC development (Ashman 1999). Not only does SCF drive MC homing, proliferation and differentiation but also MC survival, migration and functional activation. Under various experimental conditions, SCF is chemotactic for MCs and their progenitors. For example, local treatment of mice with SCF can induce marked local increase in MC number, reflecting both enhanced recruitment/retention and/or maturation of MC precursors and proliferation of more mature MCs (Tsai et al.

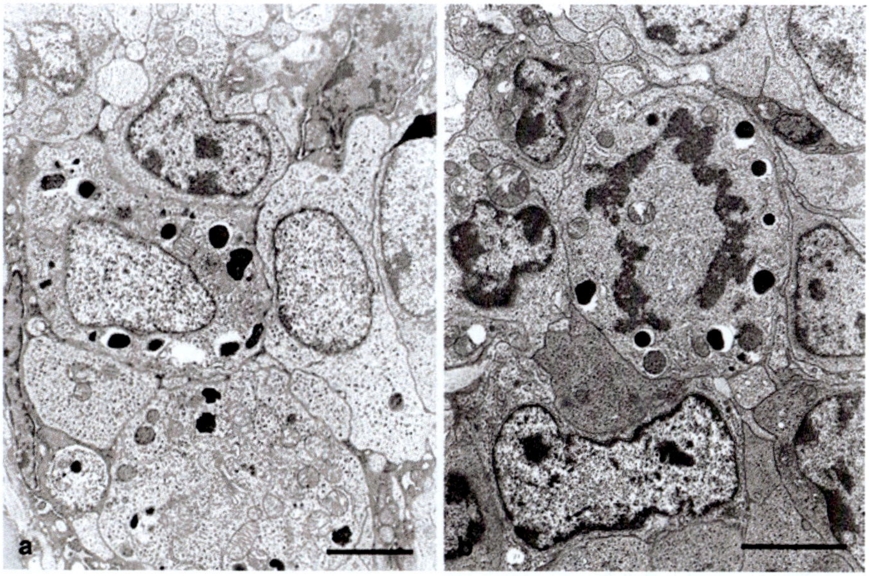


**Fig. 2.8** Transmission electron microscopy of the thymus in a day 16 chicken embryo. In (**a**) two mast cells are seen in the thymic medulla; in (**b**) a mast cell in mitosis is depicted. Bars: 2 μm in (**a**), 3 μm in (**b**). (Reproduced from Crivellato et al. 2005a)

1991). The importance of SCF as a MC growth factor is underlined by the fact—as we have said previously—that mice with certain loss-of-function mutations affecting either SCF or its receptor KIT are devoid of MCs. Indeed, lack of expression of a functional KIT receptor due to spontaneous mutation in both copies of *Kit*, as it occurs in genetically MC-deficient WBB6F1-$Kit^{W}$-$Kit^{W-v}$ mice ($W/W^{v}$ mice), results in a virtual absence of tissue MCs (Kitamura et al. 1978). $Kit^{W}$ contains a point mutation that encodes a truncated KIT protein, which lacks the transmembrane domain and is therefore not expressed on the cell surface; $Kit^{W-v}$ encodes a mutation in the KIT tyrosine kinase domain that markedly decreases the kinase activity of the receptor. A *Kit*-mutant mouse has been characterized more recently, the C57BL/6-$Kit^{W-sh/W-sh}$ mice (Grimbaldeston et al. 2005; Zhou et al. 2007). $Kit^{W-sh}$ contains an inversion mutation of the transcriptional regulatory elements upstream of the *Kit* transcription start site on mouse chromosome 5 (Galli et al. 2005a). Remarkably, MCs develop in $W/W^{v}$ mice and in C57BL/6-$Kit^{W-sh/W-sh}$ mice if these mice receive bone marrow cells from normal littermates. The mechanism of SCF in human MCs has basically been identified. Binding of SCF induces autophosphorylation of KIT and subsequent activation of several signalling molecules including PI3K and mitogen-activated protein kinase (Lorentz et al. 2002).

The specific localization of MCs in homing tissues is dependent on their interaction with extracellular matrix proteins. Cell adherence to extracellular matrix proteins is mediated by specific cell adhesion receptors, mainly cell surface receptors of the integrin family. An important integrin expressed by human MC progenitors is the α4β1 integrin, which regulates their adhesion to activated endothelial cells (Boyce et al. 2002). Mucosal MCs also possess β7 integrin, mediating the tissue specific homing of intestinal MC progenitors (Gurish et al. 2001). In the mouse, large numbers of MC-committed precursors are constitutively recruited in the small intestine by a mechanism involving the α4β7 integrin (Gurish and Austen 2001). This integrin expressed on the surface of MC precursors binds to the "mucosal address in cell adhesion molecule-1" (MAdCAM-1) and to "vascular cell adhesion molecule-1" (VCAM-1) as endothelial counterligands for this integrin. Inflammation-induced recruitment of human MC progenitors to the lungs requires both α4β7 and α4β1, implicating organ-specific control of MC progenitors influx (Gurish and Boyce 2006). Remarkably, dendritic cell expression of the transcription factor T-bet, which controls interferon (IFN)-γ production and Th1 cell differentiation from $CD4^{+}$ T cells, regulates MC progenitor homing to mucosal tissue. Indeed, homing of MC progenitors to the lung or small intestine in T-bet(-/-) mice is reduced (Alcaide et al. 2007). In addition, chemokine receptors expressed by MC progenitors are most likely involved in directing the progenitors from the circulation into the tissues where they mature. Human MC precursors derived *in vitro* from cord blood express a set of chemokine receptors including CXCR2, CCR3, CXCR4, and CCR5, and respond to the corresponding ligands *in vitro* (Ochi et al. 1999). Notch receptor-mediated signalling is involved in MC differentiation and homing. Recently, it has been recognized that Notch2 signalling in MCs is required for proper localization of intestinal MCs, and is critical for MC host-pathogen interface in the small intestine (Sakata-Yanagimoto et al. 2011).

Besides SCF, IL-3 plays in rodents an additional foundamental role in MC development (Lantz et al. 1998). By contrast, the role of IL-3 on the development of human MCs is controversial (Saito 2005). Other cytokines and growth factors which regulate MC development and differentiation include IL-4, IL-9, IL-10, transforming growth factor beta (TGF-β) and nerve growth factor (NGF) (Okayama and Kawakami 2006). IL-4 does not affect MCs by itself, but acts synergistically with SCF in the control of MC survival, proliferation as well as IgE-dependent mediator release (Bischoff et al. 1999). The IL-4 priming of human MCs for enhanced proliferation and mediator release is associated with an increased activity of extracellular signal-regulated kinase (ERK) and c-Fos (Lorentz et al. 2005). Remarkably, IL-4 changes the cytokine profile released by mature MCs by reducing proinflammatory cytokines such as TNF-α and IL-6 and in turn enhancing Th2 cytokines such as IL-5 and IL-13 (Lorentz et al. 2000). IL-9 alone does not have any effect on bone marrow-derived MC (BMMC) proliferation but mouse BMMCs undergo phenotypic changes in the presence of IL-9 in combination with SCF that consist in the acquisition of a mucosal MC phenotype (namely, accumulation of mMCP (monocyte chemoattractant protein)-1 and mMCP-2 transcripts) (Okayama and Kawakami 2006). In human systems, IL-9 and IL-5 stimulate SCF-mediated proliferation of MCs from cord blood cells, bone marrow and peripheral cells (Matsuzawa et al. 2003). IL-10 alone does not support the growth and differentiation of MC progenitors. However, when combined with IL-3 or IL-4, IL-10 enhances their growth. NGF promotes proliferation and differentiation of mouse BMMCs in the presence of IL-3 (Matsuda et al. 1991). NGF does not affect human MC survival but in combination with SCF it synergistically suppresses MC apoptosis (Kanbe et al. 2000).

Committed progenitors are supposed to populate peripheral tissues functioning as a local reservoir. These undifferentiated but committed progenitors do not develop into mature MCs unless adequate inflammatory stimuli ensue. In the adult mouse, for instance, it has been shown that the mucosa of the intestine contains the largest peripheral pool of these committed progenitors (Guy-Grand et al. 1984). Mature MCs can be very long-lived cells, surviving in some cases for years, and can retain their ability to proliferate under certain conditions (Galli and Lantz 1999).

### *2.2.3 Mast Cell Organ and Tissue Distribution*

MCs are virtually ubiquitous cells. They reside in almost all of the major organs and tissues of the body. Normally, they localize in proximity to surfaces that interface the external environment, which are common portals for pathogen, allergen and toxin entry. Thus, MCs are strategically located at host/environment interfaces like the skin, airways and gastrointestinal and urogenital tracts. MCs are likely to be among the first inflammatory cells to interact with invading microorganisms and initiate immune responses (Metz et al. 2008). In this perspective, MCs have been shown to increase in inflammatory reactions of the skin as well as the intestinal and respiratory tracts in different species, such as man, rodents, and chickens

(Rose et al. 1980; Morris et al. 2004). MCs also populate the connective tissues, particularly in association with structures such as blood vessels, lymphatic vessels and nerves, in a position which make them key elements in processes like wound healing, tissue regeneration and remodelling after injury, fibrosis and angiogenesis (Gonzalez et al. 1999; Weller et al. 2006; Grimbaldeston et al. 2007). MCs also reside in proximity to vulnerable spaces such as the peritoneum and the joint cavities. MCs are not found in avascular tissues such as mineralized bone, cartilage and the cornea. In the human body, they collectively comprise a substantial cell population. It has been estimated that if all human tissue MCs were amassed together in a single organ, it would equal the size of a normal spleen (Sayed et al. 2008). This indicates a strong selective pressure in maintaining this cell type throughout the evolutionary scale (Crivellato and Ribatti 2010b).

In humans, MCs express farly different phenotypes according to their homing site. For instance, the vast majority of MCs in the alveolar interstitium contains tryptase and only little chymase (Irani et al. 1989). However, chymase-expressing MCs are a major constituent of MC populations of the pleura and of the airway and vessel wall (Martin et al. 1992; Andersson et al. 2009). A major increase of chymase-positive MCs was noted in the adventitia of small pulmonary arteries in patients who died of asthma (Shiang et al. 2009). In addition, chymase-expressing MCs are also prominent in scarring lung diseases, like interstitial lung disease (Edwards et al. 2005).

In the human intestinal mucosa, MCs consist of approximately 2–3% of the inflammatory cell infiltrate localized in healthy subjects (Bischoff 2009). $MCs_T$ prevail in the lamina propria whilst $MCs_{TC}$ are prevalent in the submucosa. These figures change during pathological conditions. In the course of intestinal diseases, such as food allergy and parasitosis, this amount can augment up to tenfold. $MC_C$ are located mainly in the submucosa of the small intestine and colonic mucosa (Irani and Schwartz 1994). MCs are regarded as key elements of the gut-associated lymphoid tissue (GALT) and participate to several aspects of mucosal defence. In addition, intestinal MCs perform regulatory functions to maintain tissue homeostasis such as the control of the intestinal barrier. MCs density in the intestinal lamina propria has been linked to the maintenance of normal villus architecture. Low mast cell density in the human duodenal mucosa from chronic inflammatory duodenal bowel disorders is associated with defective villous architecture (Crivellato et al. 2003b). A structural interaction among crypt epithelial cells, subepithelial myofibroblasts and pericryptal MCs has been identified in the human small bowel mucosa with potential functional implications in enterocyte proliferation and differentiation (Crivellato et al. 2005b; Fig. 2.9).

In human skin, MCs are preferentially situated in the most superficial layers (more than 80 MCs/mm$^2$ in the papillary dermis) where up to tenfold more MCs are to be found as compared with the subcutis (Weber et al. 2003). Remarkably, MC numbers are highest at peripheral skin sites, such as the chin and the nose (around 50 MCs/mm$^2$), and lowest at central skin sites, such as the abdomen (around 20 MCs/mm$^2$). Thus, healthy human skin exhibits a proximal/distal and a central/peripheral MC gradient. Recently, the human skin has been demonstrated to be an

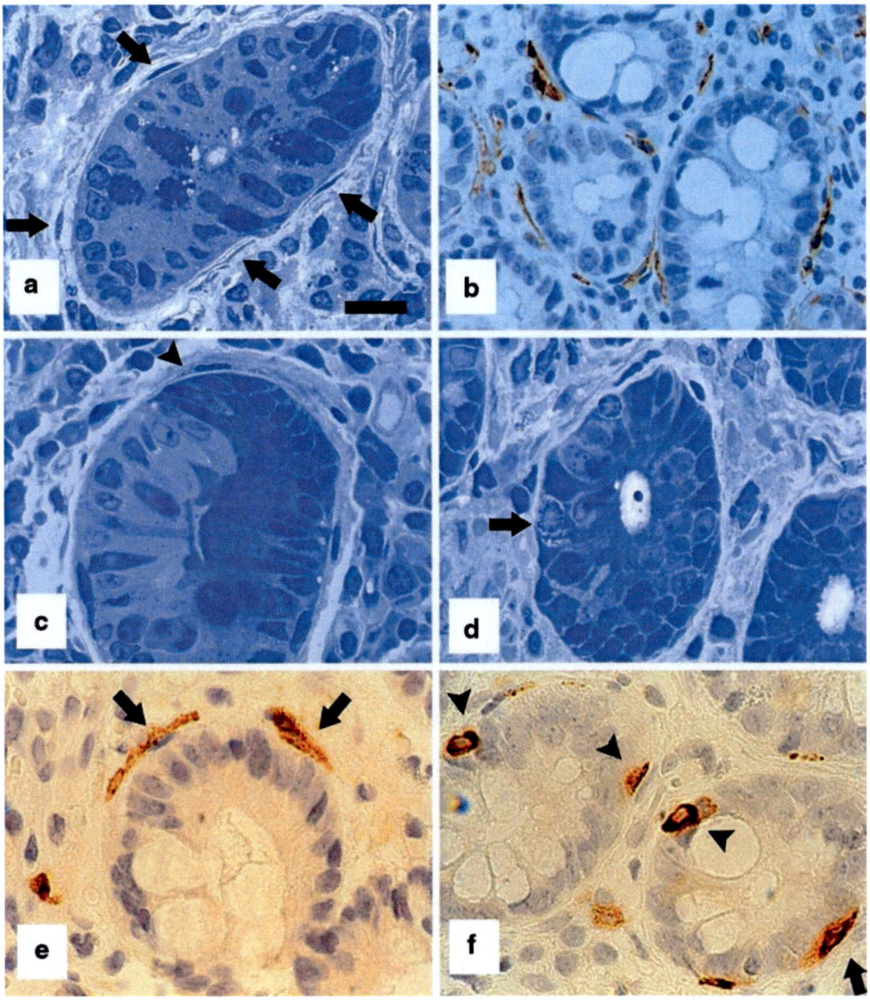


**Fig. 2.9** Microscopic anatomy of the crypt epithelium and pericryptal lamina propria in the human small bowel mucosa. Each crypt is surrounded by a sheath of intestinal subepithelial myofibroblasts (*arrows* in **a**), which express α-smooth muscle actin immunoreactivity (**b**). Mast cells with a fibroblast-like profile reside in the lamina propria close to crypts (*arrowhead* in **c**, *arrows* in **e** and **f**). Other mast cells localize within the crypt epithelium (*arrow* in **d** and *arrowheads* in **f**). Mast cells in relation to crypts contain the proteolytic enzymes tryptase (**e**) and chymase (**f**). Bars: 30 µm. (Reproduced from Crivellato et al. 2005b)

extramedullary reservoir for MC precursors, which reside in the connective tissue sheath of hair follicle (Ito et al. 2010). Remarkably, a local regulatory loop between corticotropin-releasing hormone (CRH) and SCF signalling has been identified which promotes generation of mature MCs from MC precursors in the hair follicle.

The uterus is an important site of MC homing. MCs localize to the endometrium, myometrium and cervix (Cabanillas-Saez et al. 2002). Uterine MCs are morphologically similar to skin and lung MCs (Massey et al. 1991). In the mouse, MCs are implicated in the process of angiogenesis in the cervix during pregnancy (Varayoud et al. 2004). Remarkably, the activity of these MCs as well as their number and histamine content are regulated by female reproductive hormones and increase during the second half of murine pregnancy reaching to maximum by the end of the gestational period (Rudolph et al. 1997). Notably, MC content of histamine comes to normal after delivery. By the end of pregnancy, uterine MCs release histamine, serotonin, prostaglandin $D_2$ ($PGD_2$) and leukotrienes (LTs), which stimulate uterine contraction (Bytautiene et al. 2003). It has been shown, indeed, that uterine MCs play an important role in parturition by effecting uterine contraction induced by estrogen. Drugs such as β-adrenergic agonists or corticosteroids which stabilize MC degranulation prevent preterm labor (Martínez et al. 1999). In human uterine leiomyomas, MCs have been shown to express leptin by immunocytochemistry and leptin was partly confined to tryptase-positive granules (Ribatti et al. 2007a).

MCs are present in normal and in diseased human heart tissue. Within heart tissue, MCs lie between myocytes and in close contact with blood vessels (Patella et al. 1995; Marone et al. 1995). They are also found in the coronary adventitia. Isolated human heart MCs release preformed mediators, such as histamine and tryptase as well as newly generated mediators, like $PGD_2$ and $LTC_4$ after stimulation with immunological or non-immunological stimuli. Human cardiac MCs possess FcεRI and IgE bound to the surface and receptors for the fifth component of complement (C5a), which could explain how cardiac MCs are involved in systemic and cardiac anaphylaxis. Cardiac MCs and those in human coronary arteries also play a role in the early and late stages of atherogenesis and during ischemic myocardial injury.

Recently, the white adipose tissue has gained interest in MC biology. The white adipose tissue is a heterogeneous tissue, found in various locations throughout the body, containing mature adipocytes and a stroma-vascular fraction. The latter component includes a large proportion of immune haematopoietic cells, among which MCs that contribute to diet-induced obesity (Liu et al. 2009). It has been demonstrated that MCs present in mice adipose tissue derive from haematopoietic stem/progenitor cells identified in the tissue (Poglio et al. 2010). Thus, adipose-derived haematopoietic stem/progenitor cells contain a precursor-cell population committed to the MC lineage, and able to efficiently home to peripheral organs such as intestine and skin, where it acquires properties of functional tissue MCs. Additionally, the white adipose tissue contains a significant MC progenitor population, suggesting that the entire MC lineage process take place in the white adipose tissue. Thus, considering the quantitative importance of the white adipose tissue in the adult organism and the increasing roles recently assigned to MCs in physiopathology, this highly specialized tissue may represent an important source of MCs in physiological and pathological situations.

MCs are well represented in the joint cavities. They constitute around 3% of cells in the immediate vicinity of the normal human synovial lining (Castor 1960). They are not observed within the synovial lining layer itself but rather populate the subsynovial loose connective tissue and adipose tissue, where they cluster near blood vessels and nerves (Nigrovic and Lee 2007). They are not seen within the cartilage and are rare in normal periarticular bone. Synovial MCs mostly belong to the $MC_{TC}$ phenotype, although MC staining for tryptase alone ($MC_T$) are also observed in varying proportion, usually in MCs found close to the synovial lining (Buckley et al. 1998). In the normal mouse joint, MCs are also localized to the synovial sublining, where they cluster around blood vessels and nerves (Nigrovic and Lee 2007).

MCs have also been identified in the central nervous system. Here, they are most numerous in the leptomeninges, thalamus and hypothalamus and in the dura mater of the spinal cord (Johnson and Krenger 1992). In particular, they have been found in the infundibulum, pineal organ, area postrema, choroid plexuses and in the leptomeninges surrounding the pineal organ and infundibulum (Dropp 1979). Occasional MCs are also seen within the supraoptic crest, the subfornical organ and the ventricles. In addition, they have often been recognized at sites directly adjacent to nerves in different peripheral districts, such as the skin, the intestinal mucosa and the submesothelial lamina of the peritoneum (Stead et al. 1987, 1989; Crivellato et al. 1991).

## 2.2.4 *Mast Cell Receptors*

MCs express on their surface a series of important receptors which drive their differentiation and functional activity (Fig. 2.10). One of the most significant is the high affinity receptor for IgE, FcεRI, which mediates interaction with IgE. In general terms, IgE response is accomplished by interaction with two structurally unrelated receptors: the high affinity receptor, FcεRI, which engages IgE with a 1,000-fold higher affinity ($K_a$ $10^{10}$ $M^{-1}$) than does the low affinity receptor FcεRII or CD23 ($K_a$ $10^7$ $M^{-1}$) (Sutton and Gould 1993). Mammalian MCs, as well as basophils, express the tetrameric $\alpha\beta\gamma_2$ form of the FcεRI on their surface (Galli et al. 2005a). The cross-linking of IgE with bivalent or multivalent antigen results in the aggregation of FcεRI, which is sufficient for initiating down-stream signal transduction events that activate MC exocytosis as well as the *de novo* synthesis and secretion of lipid mediators and cytokines (Rivera and Gilfillan 2006).

MCs, but not basophils, express the KIT receptor for SCF which represents a key feature for distinguishing between the two cell types. The expression of the tyrosine kinase KIT receptor on the surface of the MCs is very important for MC functional activity. Indeed it does not only drive terminal differentiation of the MCs but has also other important roles in regulating MC biology, such as survival, activation and degranulation of mature MCs. As we have previously seen, the importance of SCF as a MC growth factor is underlined by the fact that mice with certain loss-of-function mutations affecting either SCF or its receptor KIT are devoid of MCs. Indeed, lack of expression of a functional KIT receptor due to spontaneous mutation in both copies of *Kit*, as it occurs in genetically MC-deficient WBB6F1-$Kit^{W}$-$Kit^{W-v}$ mice ($W/W^{v}$ mice), results in a virtual absence of tissue MCs (Kitamura et al. 1978).

Besides the FcεRI and KIT receptors, MCs express a large array of adhesion molecules and chemotactic factor receptors. Adhesion molecules on both progeni-

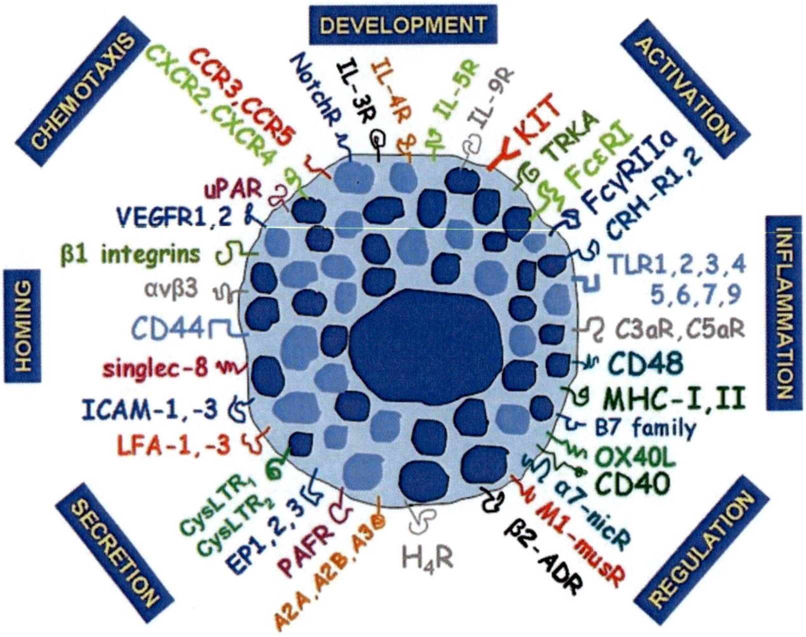


**Fig. 2.10** Human and mouse mast cells are endowed with a great number of cell surface receptors which mediate different physiological functions typical of these cells. Here, for the sake of clarity, mast cell receptors have been grouped into seven categories according to the main aspect of cell function they are believed to promote, i.e., cell development, cell activation, involvement in inflammation, cell regulation, stimulation of exocytosis, cell homing and chemotaxis. Obviously, many of these receptors may interevene in more than one functional pathway, such as the KIT receptor, which stimulates mast cell development, activation and secretion, and the FcεRI, which mediates mast cell activation and secretion

tors and mature MCs are certainly important factors controlling MC homing within tissues (Vliagolftis and Metcalfe 1997) as well as MC activation. Studies with *ex vivo* MCs obtained from human tissues demonstrate surface expression of β1 integrins such as VLA-3, VLA-4, VLA-5 and αvβ3 integrin (Valent and Bettelheim 1992; Columbo et al. 1995). The natural ligands of VLA-3, VLA-4 and VLA-5 and αvβ3 integrin are laminin, type I collagen and fibronectin; fibronectin and VCAM-1; fibronectin; and vitronectin, fibronectin, thrombospondin and fibrinogen, respectively. It has been reported that β1 integrins are involved in MC activation, upregulation of cytokine expression and survival (Ra et al. 1994). Studies in humans show that MCs from uterus and lung express the β1 integrins α4β1 and α5β1, known as receptors for fibronectin, and that skin MCs express α3β1 and adhere to fibronectin and laminin (Columbo et al. 1995). Besides these integrins, human intestinal MCs express α2β1 integrin (Lorentz et al. 2002). BMMCs have been shown to express α4, α5 and α6 integrins (Fehlner-Gardiner et al. 1999). As far as non-integrin adhesion molecules are concerned, human MCs have been reported to express low levels of intracellular adhesion molecules 1 and 3 (ICAM-1, ICAM-3) as well as leukocyte function-associated antigen-1 and 3 (LFA-1, LFA-3) (Bochner and Schleimer 2001). Additional adhesion molecules expressed by MCs are CD44, a hyaluronic acid receptor, and singlec-8, a molecule which binds to sialic acid moieties (Bochner and Schleimer 2001). MCs express surface receptors that depend on their anatomical location and the stage of differentiation and activation. Thyroid MCs, for instance, express thyroid hormone receptors (Catini and Legnaioli 1992) and genital tract MCs are responsive to estrogen and luteinizing hormones (Maurer et al. 2003). Human MCs express androgen receptors but treatment with testosterone exerts no influence on IgE-independent MC degranulation elicited by neuromuscular blocking agents (Chen et al. 2010).

In the resting state, MCs express the activating IgG receptor FcγRIIa (CD32a) and, upon interferon gamma (IFN-γ) activation, the high affinity activating FcγRI (CD64). MCs might also express the complement receptors C3aR and C5aR, receptors for various interleukins, such as IL-3R, IL-4R, IL-5R, IL-9R, IL-10R, IL-15R, growth factors and chemokines, among others the NGF receptor TRKA, GM-CSFR, INF-γR, and the receptors for chemokines CCR3, CCR5, CXCR2, CXCR4 (Nickel et al. 1999; Romagnani et al. 1999). MCs express distinct IL-15Rα isoforms which modulate MC-dependent innate immune response by fine-tuning defined MC protease activities (Orinska et al. 2007). Expression of the receptors for IL-1 family molecules, specifically, IL-1R1, IL-18R and T1/ST2, are detectable intracellularly in human umbilical cord blood-derived MCs by flow cytometry, but is scarcely detectable on the cells' surface (Iikura et al. 2007). However, IL-1β, IL-18 or IL-33 induce phosphorylation of Erk, p38 and JNK in naïve human umbilical cord blood-derived MCs, and IL-33 or IL-1β, but not IL-18, enhance the survival of naïve human umbilical cord blood-derived MCs and promote their adhesion to fibronectin. IL-33 or IL-1β also induce IL-8 and IL-13 production in naïve human umbilical cord blood-derived MCs, and enhance production of these cytokines in IgE/anti-IgE-stimulated human umbilical cord blood-derived MCs, without enhancing secretion of either $PGD_2$ or histamine. Moreover, IL-33-mediated IL-8 production by human umbilical cord blood-derived MCs is markedly reduced by the p38 MAPK inhibitor, SB203580. In contrast to findings with mouse MCs, IL-18 neither induces nor enhances secretion of the mediators $PGD_2$ or histamine by human umbilical cord blood-derived MCs (Iikura et al. 2007).

MCs release histamine but they are also sensible to histamine activity through surface receptors. MCs have been shown to express H2R, H3R and H4R receptors (Jutel et al. 2009). H2R negatively regulates the release of histamine on MCs. The control of MCs by histamine acting on H3R involves neuropeptide-containing nerves and might be related to a local neuron-MC feedback loop controlling neurogenic inflammation (Dimitriadou et al. 1994). H4R has been found on MCs. Histamine released by basophils and MCs themselves may potentiate MCs chemotaxis on the inflammatory setting by interacting with the H4 receptor (Hofstra et al. 2003). In addition, MCs express the $CysLTR_1$ and $CysLTR_2$ receptors for leukotrienes. Thus, cysteinyl leukotrienes produced by immunologically activated

MCs may exert through an autocrine loop a variety of responses by activating the receptors $CysLTR_1$ and $CysLTR_2$ (Mellor et al. 2001). IL-4 and $LTC_4$, secreted by MCs and basophils, upregulate the expression of $CysLTR_1$ and stimulate $LTC_4$ and cytokine production by human MCs. MCs express the EP(2) receptor for $PGE_2$. Human MCs are a potent source of vascular endothelial growth factor (VEGF) through activation of the EP(2) receptor (Abdel-Majid et al. 2004). MCs have recently been shown to express a platelet-activating factor (PAF) receptor (Kajiwara et al. 2010). PAF induces histamine release from human lung MCs, a mechanism which provides an amplification loop for MC activation in the generation of anaphylaxis. PAF receptor was not found in skin MCs. This receptor is responsible for a rapid PAF-induced MC degranulation.

MCs also express receptors for the purine nucleoside adenosine. This molecule is produced in high concentration during tumour growth and it has been implicated in promoting angiogenesis. Interestingly, the human MC line HMC-1 expresses A2A, A2B and A3 adenosine receptors (Feoktistov et al. 2003). Selective stimulation of the A2B receptor increases VEGF and IL-8 secretion by HMC-1 MCs whilst the A3 receptor increases the expression of angiopoietin (ang)-2 mRNA. Thus, adenosine acts in a functional fashion to promote tumour angiogenesis by a cooperative paracrine mechanism involving A2B and A3 receptors on infiltrating MCs that, in turn, secrete angiogenic factors. Interestingly, selective release of VEGF by human MCs is regulated by corticotropin releasing hormone (CRH), which implies the presence of a CRH receptor on MC surface (Cao et al. 2006). In addition, $PGE_2$ dose-dependently induces primary MCs to release the proangiogenic chemokine MCP-1 through engagement of EP(1) and EP(3) receptors (Nakayama et al. 2006). Remarkably, MCP-1 is detected by immunoelectron microscopy within MCs at a cytoplasmic location distinct from the secretory granules, which implies an exocytosis-independent mode of molecule extrusion.

As far as CRH receptor is concerned, two CRH receptor subtypes, CRH-R1 and CRH-R2, have actually been detected on subepithelial MCs in the human colon (Wallon and Söderholm 2009) and CRH-R1 has been found in cutaneous MCs from a patient with urticaria pigmentosa (Theoharides et al. 2009). These receptors may explain exacerbations of digestive and cutaneous symptoms in allergic and atopic subjects or may affect the course of ulcerative colitis and irritable bowel syndrome during persistent psychological stress, which implies CRH release from the hypothalamus. Blood-brain-barrier permeability and multiple sclerosis appear to worsen in response to acute stress that leads to the local release of CRH, which activates brain MCs to selectively release IL-6, IL-8 and VEGF. In addition, acute stress shortens the time of onset of experimental allergic encephalomyelitis (EAE) that does not develop in MC-deficient $W/W^v$ mice or $CRH^{-/-}$ mice (Theoharides and Konstantinidou 2007).

MCs express adrenoreceptors on their surface. This is most important for the control of asthmatic symptoms. The human lung MC is a crucial effector cell in the mediation of asthma. Activation of MCs by allergens, and other insults, leads to the elaboration of a wide variety of autacoids that cause bronchoconstriction and promote inflammation. Of the drugs that are used to treat asthma, only bronchodilator β2-adrenoceptor agonists are effective at inhibiting the elaboration of mediators from MCs. Both short- and long-acting β2-adrenoceptor agonists are effective inhibitors of MCs. Human lung MCs express a homogeneous population of β2-adrenoceptors (Kay and Peachell 2005). The β2-adrenoceptors agonists isoprenaline and salbutamol inhibited anti-IgE-induced release of histamine, $PGD_2$ and $LTC_4$ from human peripheral blood-derived MCs in a dose-dependent manner whilst the selective β3-adrenoceptor agonist BRL-37344 failed to affect anti-IgE-induced histamine release and the β1-adrenoceptor antagonist atenolol did not have any effect (Wang and Lau 2006). B2-adrenoceptors in MCs act via G protein coupled with the receptor. In addition, MCs express cholinergic receptors. For instance, electrical vagal stimulation was observed to induce gastric mucosal MC degranulation in experimental animals. Expression of mRNA of nicotinic acetylcholine receptors-α4, -α7, and -β2 subunits were detected in mucosal-type murine BMMCs (Kageyama-Yahara et al. 2008). IgE-induced degranulation of these cells is negatively regulated via nicotinic acetylcholine receptors, in particular the nicotinic acetylcholine receptors-α7 subunit. Interestingly, MCs in atopic dermatitis but not in healthy skin showed nicotinic acetylcholine receptors-α3 and -α5 subunit immunoreactivity (Kindt et al. 2008). As for muscarinic acetylcholine receptors, a species difference exists in the cholinergic control of histamine release between human and rat airways. In human airways, muscarinic receptors most likely of the M1 subtype are involved in the inhibitory control of MC function, whereas such an inhibitory pathway does not exist in the rat trachea (Reinheimer et al. 2000).

MCs have increasingly been implicated in promoting innate immunity against pathogen invasion. In this setting, MC activation can be elicited by diverse mechanisms that include signalling through Toll-like receptors (TLRs), receptors for complement components and receptors for endogenous peptides. MCs also express adhesion molecules acting as parasite, bacterial and virus receptors (for instance the CD48) (Gilfillan and Tkaczyk 2006). Mammalian TLRs have essential roles in the direct recognition of infectious agents, initiating signalling through NF-kB leading to the initiation of both innate and adaptive immune responses (Leulier and Lemaitre 2008). Human MCs have been shown to express TLR1, TLR2, TLR3, TLR4, TLR5, TLR6, TLR7, TLR8, and TLR9 under certain conditions (Hofmann and Abraham 2009). TLR2, for instance, has been found to recognize and respond to several pathogen-associated molecular patterns, including peptidoglycan, lipoproteins, and lipotheicoic acid (Schwandner et al. 1999). MCs stimulated through TLRs express "surgical" cytokine responses consisting mainly of IL-4, IL-6, IL-8 and TNF-α secretion (Bachelet and Levi-Schaffer 2007). For instance, when TLR2 or TLR4 are activated, human MCs release TNF-α, IL-5, IL-10, and IL-13 which facilitate the recruitment of immune cells to the site of infection (Varadaradjalou et al. 2003). TLR2 and TLR4 activation stimulates production of IL-6 and TNF-α by cultured mouse MCs without affecting degranulation, arachidonic acid production, or calcium mobilization (Quiao et al. 2006).

MCs may also be activated by complement receptor (C3aR, C5aR CR2, CR4, C1qR) and release proinflammatory and chemotactic mediators (Marshall 2004; Gilfillan and Tkaczyk 2006). The complement system is a complex system of proteins that interact in a proteolytic cascade, leading to pathogen clearance in the serum. It has a more ancient origin than acquired immunity. Besides expressing on their surface major histocompatibility complex (MHC) class I and class II molecules under certain conditions, MCs represent a rich source of co-stimulatory activity by expressing also molecules of the B7 family (ICOS-L, PD-L1, PD-L2, CD80, CD86), members of the TNF/TNF receptor families (OX40L, CD153, Fas, 4-1BB), CD28 and CD40 ligand (Nakae et al. 2006; Hershko and Rivera 2010). Recently, it has been recognized that mouse BMMCs or mouse peritoneal MCs constitutively express Notch1 and Notch2 proteins on the cell surface. It has also been shown that Delta-like 1 (Dll1)/Notch signalling induces the expression of MHC-II and upregulates the expression level of the co-stimulatory molecule OX40L on the surface of the MCs. Dll1/Notch signalling augments FcεRI-mediated IL-4, IL-6, IL-13, and TNF production by murine BMMCs. Dll1-stimulated MHC-II$^+$OX40L$^{hi}$ murine BMMCs promote proliferation of naive CD4$^+$ T cells and their differentiation into Th2 cells producing IL-4, IL-5, IL-10, and IL-13 (Nakano et al. 2009). During the *in vitro* ingestion of a pathogenic *E. coli* by human MCs, production of TNF-α, CC chemokine ligands (CCL-1/I-309, CCL-19/MIP (macrophage inflammatory protein)-3β, and CCL-18/MIP-4), the integrin CD49d, and CD80 are upregulated. Remarkably, coincubation of human MCs with *E. coli* downregulates FcεRI expression and FcεRI-mediated MC degranulation (Kulka et al. 2004).

MCs also express the urokinase plasminogen activator receptor (uPAR) for the urokinase plasminogen activator (uPA), which may be related to the specific pro-angiogenic function of MCs, urokinase being an enzyme involved in processes of tissue remodelling such as fibrinolysis, fibroblast and endothelial cell migration and local repair (Sillaber et al. 1997). uPA is a potent chemoattractant for MCs. The finding that angiogenic factors, such as fibroblast growth factor (FGF)-2, VEGF and platelet-derived endothelial cell growth factor (PD-ECGF) stimulate MC migration suggests that MCs would express surface receptors for these proangiogenic cytokines. Indeed, human lung MCs isolated *ex vivo* have been shown to express mRNA for VEGFR1 and VEGFR2 receptors (Detoraki et al. 2009). These receptors are present on the MC surface. VEGF-A(165), VEGF-B(167), VEGF-C, VEGF-D, and placental growth factor (PIGF) are able to induce MC chemotaxis. These chemotactic effects are mediated by the activation of both VEGFR-1 and VEGFR-2. It has been recently demonstrated that mouse BMMCs express functional calcitonin gene-related peptide (CGRP)-1 receptors and that their activation causes mobilization of $Ca^{2+}$ from intracellular stores and piecemeal release of mMCP-1. These findings support the hypothesis that the CGRP signalling from afferent nerves to MMCs in the gastrointestinal wall is receptor-mediated (Rychter et al. 2011).

Human MCs have recently been shown to express the death receptor TRAIL (TNF-related apoptosis inducing ligand), belonging to the TNF receptor superfamily, and the inhibitory receptors CD300a and Siglec-8 (Karra et al. 2009). These receptors might represent interesting selective targets for MC downregulation during allergic diseases, mastocytosis and other inflammatory diseases in which MCs have a central role. TRAIL is the only known functional death receptor on human MCs. Interestingly, its function is upregulated by IgE-dependent MC activation. The inhibitory receptors CD300a and Singlec-8 potently downregulate MC activation and survival *in vitro* and inhibit different IgE-mediated responses *in vivo*. Remarkably, human MCs express other inhibitory receptors, such as the FcγRIIb receptor which binds IgG complexes and the leukocyte-associated immunoglobulin-like receptor (LAIR)-1 which binds collagen (Lebbink et al. 2006).

### 2.2.5 Mast Cell Activation

The cross-linking of IgE with bivalent or multivalent antigen results in the aggregation of FcεRI, which is sufficient for initiating down-stream signal transduction events that activate cell degranulation as well as the *de novo* synthesis and secretion of lipid mediators and cytokines (Blank and Rivera 2004). IgE and its receptors are believed to have evolved as a mechanism for protection against parasites. Studies in man demonstrate that the serum presence of antigen-specific IgE correlates with acquired immunity toward certain parasitic helminth infections (Rihet et al. 1991). Mice with a targeted deletion of the IgE gene have increased worm burdens and reduced granulomatous inflammation following primary infection with *Schistosoma mansoni* (King et al. 1997). So far, IgE as well as IgG have been found exclusively in mammals. Classical IgE-mediated MC activation leads to immediate extrusion of granule-associated substances, such as histamine and proteases. Within minutes, they generate lipid-derived mediators (Brody and Metcalfe 1998). Then, MCs activation is followed also within hours, by the *de novo* synthesis of numerous cytokines and chemokines (Galli et al. 2005a).

MCs may also be activated by "alternative", IgE-independent, pathways such as aggregation of FcγRIII by IgG/antigen complexes, KIT and TLR mechanisms, exposure to chemokines, anaphylatoxins C3a and C5a, fragments of fibrinogen and fibronectin (Johnson et al. 1975; Wojtecka-Lukasik and Maslinski 1992; Prodeus et al. 1997; Gommerman et al. 2000; Marshall 2004; Fig. 2.11). Activation of MCs does not necessarily cause cell degranulation. As previously remarked, when cultured mouse MCs are activated by TLR2 and TLR4 stimulation, a production of IL-6 and TNF-α is determined which does not affect MC degranulation, arachidonic acid production, or calcium mobilization (Quiao et al. 2006). Notably, activation of MCs by TLRs leads to "surgical" release of mediators by MCs (McCurdy et al. 2003; Bachelet and Levi-Schaffer 2007). MCs may be activated by complement receptor (C3aR, C5aR CR2, CR4, C1qR) and release proinflammatory and chemotactic mediators (Marshall 2004; Gilfillan and Tkaczyk 2006). Of note, MCs can recognize, attach to, and internalize a wide variety of opsonized bacteria (Féger et al. 2002). For example, *Salmonella typhimurium* coated with the iC3b fragment of complement is recognized through CR3 on the MC membrane (Sher et al. 1979). In the cecal ligation and puncture model, it has been shown that mice deficient for the CR3b ($CR3^{-/-}$) have impaired MC activation and neutrophil recruitment, associated with reduced bacterial clearance and survival (Rosenkranz et al. 1998). Thus, MCs exploit the CR3 moiety for effecting important activation programs. Human

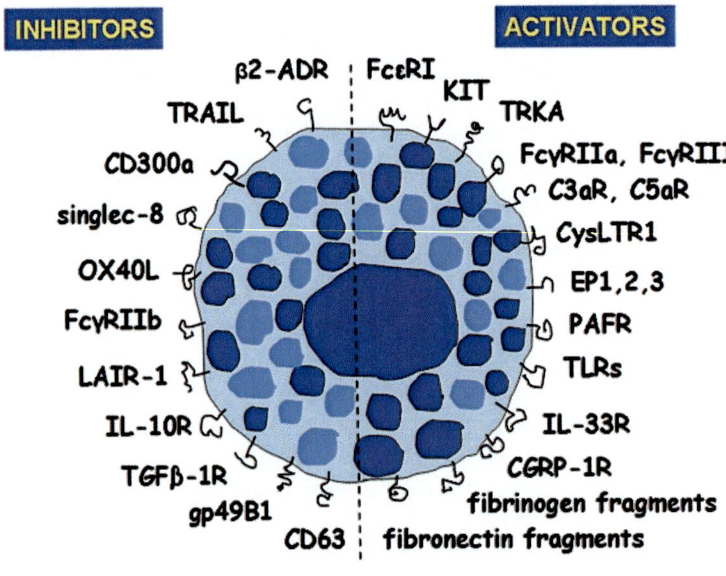


**Fig. 2.11** Mast cells express on the cell surface a series of receptors which mediate either cell activation (on the *right*) or cell inhibition (on the *left*). Activation factors promote stimulation of cell metabolism and secretion from mast cells of pre-stored and/or newly-formed molecules. The FcεRI, which mediates interaction with IgE, and the KIT receptor for stem cell factor are the most important activation receptors. Mast cells activation may be promoted by a number of "alternative" pathways, such as neuropeptide or complement stimulation pathways as well as activation through Toll-like receptor mechanisms. The list of inhibitors includes IgG complexes which bind to the FcγRIIb receptor, collagen which binds to the leukocyte-associated immunoglobulin-like receptor (LAIR)-1, the anti-inflammatory cytokines TGFβ-1 and IL-10, and several other molecules such as retinol, β2-adrenoreceptor agonists, and extracellular matrix proteins binding to CD63. Human mast cells express the death receptor TRAIL and the inhibitory receptors CD300a and Siglec-8. Treg cells directly inhibit the FcεRI-dependent mast cell degranulation through cell-cell contact involving OX40-OX40L interactions

MCs challenged to IFNγ express FcγRI at sufficient quantity to become activated for mediator release upon FcγRI aggregation (Okayama et al. 2000, 2001). This mechanism could be of relevance for IgE-independent allergic reactions as well as for nonallergic MC activation during type III hypersensitivity reactions or infections (Bischoff 2009). MCs have recently been shown to be activated through PAF receptor (Kajiwara et al. 2010). PAF induces histamine release from human lung MCs and peripheral blood-derived MCs but not skin MCs. Activation of PAF receptor-coupled $G_{\alpha i}$ leads to degranulation through phospholipase C-γ1 and phospholipase C-β2 activation in human MCs. PAF-induced degranulation is rapid, being maximal at 5 seconds, and is partially dependent on extracellular $Ca^{2+}$. These findings provide a mechanism whereby PAF mediates an amplification loop for MC activation in the generation of anaphylaxis. Another IgE-independent mode of MC activation is mediated by IL-33. IL-33 is a recently identified member of the IL-1 family of molecules, which also includes IL-1 and IL-18. IL-33 binds to the receptor, T1/ST2/IL-1R4. IL-33, like IL-1β, can induce cytokine production in human MCs even in the absence of stimuli of FcεRI aggregation (Iikura et al. 2007). These findings support the hypothesis that IL-33 may enhance MC function in allergic disorders and other settings, either in the presence or absence of co-stimulation of MCs via IgE/antigen-FcεRI signals. In the mouse, IL-33 induces IL-13 production by MCs independently of IgE-FcεRI signals (Ho et al. 2007).

As previously discussed, preformed substances stored in MC secretory granules can be released by two morphologically distinct secretory pathways, referred to as anaphylactic degranulation and piecemeal degranulation (Dvorak 2005b). Vesicle-mediated degranulation in MCs is effected by IgE-independent activation mechanisms (Theoharides et al. 2007). Instead of utilizing the high-affinity IgE receptor, FcεRI, MCs exploit such receptors as low-affinity IgG receptors FcγRIIa and FcγRIII, complement receptors (C3aR, C5aR CR2, CR4, C1qR), and Toll-like receptors (TLR-2, -3, -4, -7 and -9) (Féger et al. 2002; Marshall 2004; Gilfillan and Tkaczyk 2006). Activated MCs through these receptors selectively release mediators capable of modulating neighbouring cell functions in a manner tailored to the stimulus. It has been recently demonstrated that murine BMMCs express functional CGRP-1 receptors and that their activation causes mobilization of $Ca^{2+}$ from intracellular stores and piecemeal release of mMCP-1. These findings support the hypothesis that the CGRP signalling from afferent nerves to MMCs in the gastrointestinal wall is receptor-mediated (Rychter et al. 2011).

Polyamines mediate numerous cellular and physiological functions and recently these substances have been implicated in MC activation. MCs express antizyme inhibitor 2 (AZIN2), an activator of polyamine biosynthesis (Kanerva et al. 2009). Immunostainings show that AZIN2 is expressed in primary and neoplastic human and rodent MCs. AZIN2 localizes in the Vamp-8 positive, serotonin-containing subset of MC granules, but not in tryptase-containing granules, as revealed by double immunofluorescence stainings. Furthermore, activation of MCs induces rapid upregulation of AZIN2 expression and its redistribution, suggesting a role for AZIN2 in secretory granule exocytosis. Release of serotonin from activated MCs is polyamine-dependent whereas release of histamine and β-hexosaminidase is not, indicating a granule subtype-specific function for polyamines.

Activation of MCs leads to different patterns of biological response according to the anatomical site and the phenotypic profile of involved MCs. In comparison with MCs isolated from human skin, lung and heart, synovial MCs appear to exhibit particularly brisk responses to the neuropeptide substance P (Kopicky-Burd et al. 1988; de Paulis et al. 1996). Unlike cutaneous MCs, MCs from osteoarthritis synovium do not respond to anaphylatoxin C5a or morphine (Verbsky et al. 1996). In addition to histamine, synovial MCs obtained from osteoarthritis specimens are capable of elaborating approximately equivalent amounts of $PGD_2$ and $LTC_4$ when stimulated by IgE cross-linking, unlike MCs in the skin that synthesize $PGD_2$ preferentially (de

Paulis et al. 1996). Activation of cardiac MCs *in vitro* with anti-IgE or anti-FcεRI induces the release of preformed mediators, such as histamine, tryptase, chymase, and renin as well as the *de novo* synthesis of $LTC_4$ and $PGD_2$. C5a causes rapid release of histamine and tryptase from human cardiac MCs (Genovese et al. 2010). Remarkably, atrial natriuretic peptide is not a cardiac MC secretagogoue whereas it induces significant histamine release from peritoneal MCs (Murray et al. 2007). An important link has been identified in the heart between sensory nerves and renin-containing MCs. Substance P released from sensory nerves plays a significant role in the release of MC renin in ischemia/reperfusion and in the activation of a local cardiac renin-angiotensin system (Morrey et al. 2010). This culminates in angiotensin production, noradrenaline release, and arrhythmic cardiac dysfunction.

During the last few years, *in vitro* experiments and experiments in animal models have lead to the discovery that there are several inhibitory mechanisms operating in MCs which might counterbalance the effects of activating mediators. The list

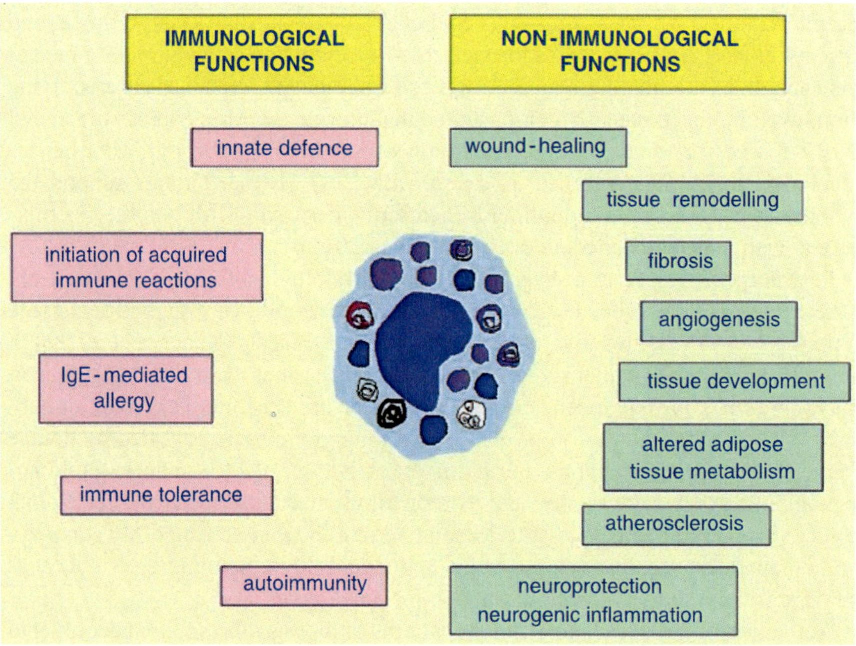


**Fig. 2.12** The broad spectrum of mast cell activities may roughly be summarized into two main categories, that is immunological and non-immunological functions. Immunological functions entailing substantial mast cell contribution are (1) the innate defence toward bacteria, protozoa and viruses, (2) the initiation and orchestration of acquired immune reactions, (3) polarization toward IgE-mediated responses, (4) the induction of immune tolerance and (5) the crucial support to autoimmunity. Non-immunological functions encompass an even greater spectrum of mast cell specializations, ranging from a substantial contribution to the mechanisms of (1) wound-healing, (2) tissue remodelling, (3) fibrosis and (4) angiogenesis, to the relevant participation to physiological and pathological events, such as (5) tissue development, (6) altered adipose tissue metabolism, (7) atherosclerosis and (8) neuroprotection and neurogenic inflammation

of inhibitors includes ligands of immunoreceptor tyrosine-based inhibition motif-containing receptors such as FcγRIIB, gp49B1, SIRPα, the human analogs LIR-5, and LILR B4 as well as the anti-inflammatory cytokines TGFβ-1 and IL-10, CD200, intracellular signal molecules like the transmembrane adaptor non-T cell activation linker (NTAL) or RabGEF1, and several other molecules such as retinol, β2-adrenoreceptor agonists, and extracellular matrix proteins binding to CD63 (Bischoff 2007).

Release of preformed and newly formed MC products from activated MCs leads to a series of profound biological effects. For the sake of clearness, the effects of MC activation may be conceptualized into two partly overlapping categories, i.e., immunological and non-immunological functions (Fig. 2.12).

## 2.2.6 *Mast Cell Immunological Functions*

MCs have increasingly been recognized as crucial effectors in both innate and adaptive immune responses. They promote immune responses against a large spectrum of pathogens. Their pivotal role in allergic disorders is well-known. In addition, these cells exert a critical role in orchestrating efficient immune responses as well as detrimental immunological functions such as autoimmunity. Thus, they are now considered to be a "linker" between innate and acquired immunity (Galli et al. 2005b).

### 2.2.6.1 Mast Cells and Innate Immune Responses

MCs are sentinel cells of the innate immunity (Fig. 2.13). This concept mostly derives from studies using MC-deficient mice. In these experimental settings, MCs have been shown to protect against bacteria, fungi, protozoa and perhaps even viruses through the release of proinflammatory and chemotactic mediators (Féger et al. 2002). Thus MCs have come to the forefront as triggers of innate immune responses against pathogens. Their preferential location to the lining surfaces of the body makes these cells an ideal first responder during microbial attack. Several recent reports indicate, indeed, that human and mouse MCs can mediate a variety of antimicrobial effects following activation upon contacts with pathogens. Although the physiological relevance of the phagocytic activity exerted by MCs remains undetermined, human and mouse MCs are capable to eliminate bacteria through an intracellular killing system—partly oxidative, partly nonoxydative—similar to that of professional phagocytes (Féger et al. 2002). MCs possess CR3 and FcγR and therefore have the capacity to recognize pathogens that have been osponized by either complement or IgG. For instance, MCs are able to recognize and kill the bacterium *Salmonella typhimurium* coated with the C3b fragment of complement and other IgG-coated bacteria (Sher et al. 1979; Talkington and Nickell 2001). These bacteria are then endocytosed via an endosome lysosome pathway. Remark-

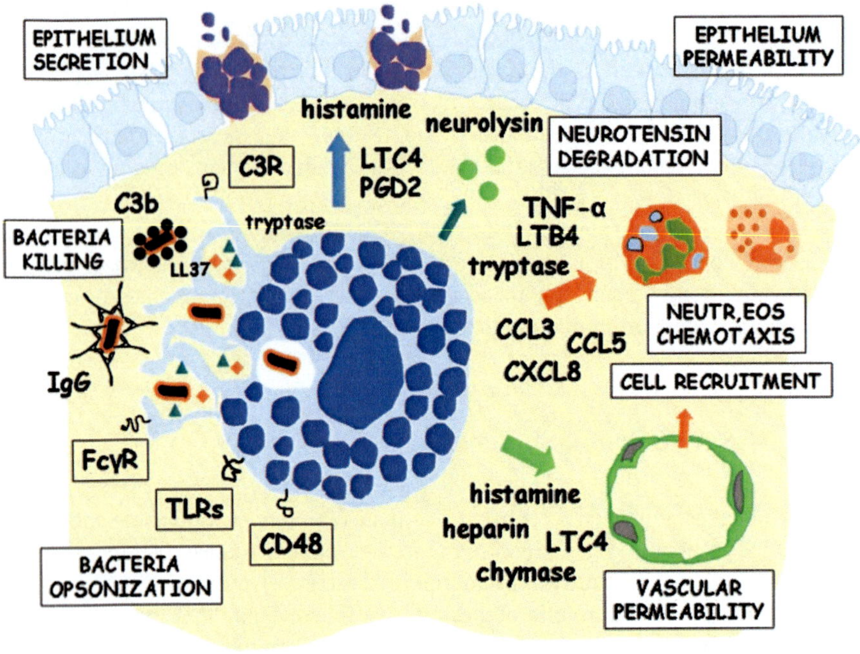


**Fig. 2.13** Mast cells are sentinel cells of the innate immunity. Human and mouse mast cells are capable to eliminate bacteria through an intracellular killing system. Mast cells possess CR3 and FcγR and therefore have the capacity to recognize pathogens that have been osponized by either complement or IgG. Mast cells have also the innate capacity to recognize pathogens through TLR or mannose receptor (CD48). Mast cells may exert bactericidal activity by an extracellular phagocytosis-independent mechanism similar to the neutrophil extracellular traps, whose major components are tryptase and the cathelicidin antimicrobial peptide LL-37. In addition, mast cells are a rich source of early-response cytokines, such as TNF-α and IL-4, that can rapidly recruit inflammatory cells at the site of pathogen entry. Other mast cell-derived products, such as leukotriene $B_4$, tryptase and chemokines such as CCL3, CCL5 and CXCL8 contribute to the influx of neutrophils. Mast cells may decrease neurotensin-induced hypotension as well as sepsis-related mortality by degrading neurotensin through the protease neurolysin. Mast cells also release vasoactive mediators, such as histamine, heparin, leukotriene $C_4$ and chymase, which induce vascular permeability. Epithelium permeability and secretion at the site of bacterial entry is stimulated by mast cell products such as histamine, leukotriene $C_4$ and prostaglandin $D_2$

ably, MCs have also the innate capacity to recognize pathogens in the absence of opsonins through TLR or mannose receptor (CD48) (Marshall 2004). Recent data suggest that MCs may exert bactericidal activity by an extracellular phagocytosis-independent mechanism, which consists in the production of extracellular structures similar to the neutrophil extracellular traps (NETs). The major components of these MC extracellular traps are DNA, histones, and granule proteins such as tryptase and the cathelicidin antimicrobial peptide LL-37 (von Köckritz-Blickwede et al. 2008). Interestingly, it has been recognized that MCs themselves may participate to the formation of granulomatous lesions caused by parasitic infection in rats (McHardy et al. 1993). In addition, the recruitment of neutrophils and subsequent macrophages elicited by skin MCs in the mice has been found to be of critical importance for the induction and development of cutaneous granuloma formation, which is crucial for the successful containment and elimination of many intracellular pathogens (von Stebut et al. 2003).

Besides their potential capacity to clear bacteria through endocytotic pathway, MCs are decisive in initiating the immune and inflammatory responses of the host to the invading pathogens. MCs, indeed, are a rich source of early-response cytokines, such as TNF-α and IL-4, that can rapidly recruit inflammatory cells at the site of pathogen entry (Galli et al. 2005a). TNF-α in particular, is a pivotal molecule in the defensive mechanisms initiated by MCs. MCs release TNF-α stored in secretory granules after incubation with bacteria both *in vitro* and *in vivo* (Echternacher et al. 1996; Malaviya et al. 1996). In mutant $W/W^v$ mice, the absence of MCs leads to a defective innate immune response against bacteria. In a model of acute septic peritonitis by cecal ligation and puncture, $W/W^v$ mice exhibited a dramatically increased mortality compared with the wild-type mice (Echternacher et al. 1996). Remarkably, the adoptive transfer of MCs to the peritoneum protected the MC-deficient mice from the lethal effects of cecal ligation and puncture. Similarly, MC-deficient $W/W^v$ mice are less protected against experimentally induced lung enterobacterial infections than MC-sufficient or MC-reconstructed $W/W^v$ mice (Malaviya et al. 1996). The impaired killing of bacteria in MC-deficient mice was directly correlated with reduced neutrophil infiltration in lungs, partly as a result of lower levels of the MC-derived chemotactic TNF-α in these mice. Indeed, TNF-α-deficient mice have increased mortality in the cecal ligation and puncture model compared with wild-type mice (Maurer et al. 1998). Other MC-stored substances are endowed with the capacity to generate important defensive responses. Recent evidence indicate that mMCP-2, a mouse MC serine protease of the chymase type, can contribute to neutrophil recruitment and host survival during cecal ligation and puncture in mice (Orinska et al. 2007). Interestingly, this protease is inhibited by IL-15, which is constitutively expressed and can be induced in MCs themselves (Orinska et al. 2007). In a mouse model of sepsis, it has been shown that MCs may decrease neurotensin-induced hypotension as well as sepsis-related mortality by degrading neurotensin through the protease neurolysin (Piliponsky et al. 2008). Remarkably, other MC-derived products, such as $LTB_4$, human tryptase βI, MIP-1α (CCL3), MIP-1β, MIP-2, MCP-1, RANTES (regulated upon activation, normal T-cell expressed and secreted) (CCL5), and IL-8 (CXCL8) appear also to contribute to the influx of neutrophils induced by activated MCs (Yamazaki et al. 1998; Féger et al. 2002).

In some experimental settings, MCs have been shown to exert detrimental effects to the host during bacterial infections by excessive or inappropriate release of inflammatory mediators leading to harmful outcome. For instance, there are indications that Shiga toxin produced by *Shigella dysenteriae* may stimulate intestinal MCs to release excessive amounts of inflammatory mediators derived from arachidonic acid metabolism, in particular $LTC_4$, leading to deleterious effects such as diarrhea and dysentery (Pulimood et al. 1998). MCs may also trigger inflammation in *Helicobacter pylori* infection, as MC accumulation in the mucosa of patients with gastritis and MC degranulation by *H. pylori* products have been described (Masini et al. 1994; Plebani et al. 1994; Nakajima et al. 1997). In addition, although MCs are capable to phagocytose and kill various opsonised bacteria, this capacity may be subverted by microbes endocytosed in nonopsonic conditions (Shin et al. 2000). In these cases, the internalized pathogen is sequestered in a MC endosomal compartment that escapes acidification and oxygen radicals entry. The net result is that the bacteria are not killed by MC but remain in the MC cytoplasm as an intracellular reservoir (Féger et al. 2002)

#### 2.2.6.2 Mast Cells and Adaptive Immune Responses

In addition to activating the innate immune system during infections, MCs have recently been recognized to exert a profound role in adaptive immunity (Galli et al. 2005b; Sayed and Brown 2007; Frossi et al. 2010) (Fig. 2.14). These cells, indeed, have been shown to be capable to influence the outcome of both physiological and pathological T cell responses. MC involvement in adaptive immune responses is broad: (1) they coordinate adaptive immune responses to pathogens; (2) contribute to the initiation of the primary immune responses to allergens; (3) amplify exacerbations of allergic diseases; (4) exert important role in generating immune tolerance; (5) primarily affect certain autoimmune diseases.

MCs may contribute to optimal initiation of acquired immunity by orchestrating migration, maturation and function of dendritic cells and by interacting with T and B cells (Hershko and Rivera 2010). MCs promote dendritic cell migration mainly by releasing prestored TNF-α at the site of infection. TNF-α, in turn, facilitates migration of dendritic cells to lymph nodes through upregulation of CCR7, a dendritic cell chemokine important for dendritic cell homing to lymph nodes (Yamazaki et al. 1998; Suto et al. 2006). TNF-α also induces maturation of dendritic cells in so far as they are induced to express MHC class II moieties and co-stimulatory molecules thereby becoming an antigen presenting cell for T cells (Ritter et al. 2003). MC-derived TNF-α drained at distance to local lymph nodes also express the capacity to retain lymphocytes circulating from the blood in the nodes (Young et al. 2000). For instance, it has been reported that TNF-α released from MCs contributes substantially to T cell recruitment to the draining lymph nodes in an experimental infection model with *Escherichia coli* (McLachlan et al. 2003). In addition, TNF-α released by MCs upon FcεRI engagement can cause T cell proliferation and cytokine production (Nakae et al. 2005, 2006). MCs may recruit effector T cells also through secretion of histamine, chemokines and $LTB_4$ (Ott et al. 2003; Demeure et al. 2005; Maurer et al. 2006; Jawdat et al. 2006). MCs may also influence the polarity of T cell responses. Activated MCs release Th2 polarizing cytokines, such as IL-4, IL-10 and IL-13, which induce stimulated naïve $CD4^+$ cells to become Th2 cells once activated in the lymph node. These Th2 cells induce strong humoral immune responses that are protective against pathogens (Stelekati et al. 2007; McLachlan et al. 2008). Upon TLR3 engagement, MCs may also express important regula-tory

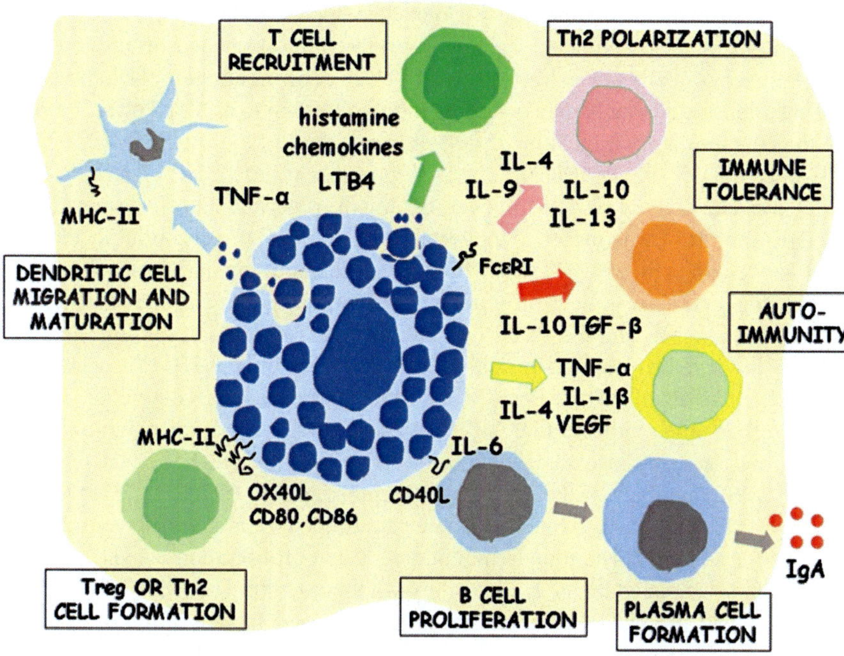


**Fig. 2.14** This schematic drawing depicts the multifarious effects exerted by mast cells in the context of adaptive immunity. Mast cells regulate immune cells through released mediators or cell-to-cell contacts. Migration and maturation of dendritic cells is mainly regulated by TNF-α, which also induces expression of MHC class II moieties and co-stimulatory molecules. TNF-α released by MCs upon FcεRI engagement can cause T-cell recruitment, proliferation and cytokine production. T cell recruitment is also effected by histamine, some chemokines and leukotriene B4. Th2 polarization is mediated by interleukines, like IL-4, IL-9, IL-10 and IL-13 released by mast cells upon FcεRI engagement. Mast cells also induce B cell proliferation and plasma cell formation through cell-cell contact, IL-6 and CD40L stimulation. Mast cells express MHC class II in limited circumstances. These mast cells may express CD80 and CD86 co-stimulatory molecules. Induction of regulatory T cell (Treg) population is effected by MHC-II and OX40L molecules. Mast cells can promote suppression of immune reactions by producing inhibitory cytokines, such as IL-10 and TNF-α. Several lines of evidence indicate that MC may play a role in autoimmunity by producing several inflammatory mediators, notably IL-4, TNF-α, IL-1β and VEGF

functions for $CD8^+$ T cell activities both *in vivo* and *in vitro* (Orinska et al. 2005). Under certain conditions, MCs can directly activate T cells by functioning as antigen presenting cells (Frandji et al. 1996). MCs express MHC class I and MHC class II in limited circumstances. Malaviya et al. (1996) demonstrated that association of bacterial antigens and MHC class I molecules on the surface of MCs may induce $CD8^+$ T cell responses to pathogens. Recently, Kambayashi et al. (2009) provided evidence for expression of MHC class II on the surface of murine BMMCs as well as murine peritoneal MCs when activated with lipopolysaccaride (LPS) and IFN-γ. Remarkably, MHC class II-expressing MCs seemed to travel from the activation site to regional lymph nodes like dendritic cells. These MCs were also seen to express CD80 and CD86 co-stimulatory molecules. Such MHC class II-bearing MCs were demonstrated to activate T cells with preferential expansion of antigen-specific regulatory T cells (Tregs) over naïve $CD4^+$ T cells. In addition, Nakano et al. (2009) demonstrated that Notch signalling induce expression of MHC class II along with the co-stimulatory molecule OX40L on the surface of murine BMMCs. These MHC class II, OX40L BMMCs are able to promote proliferation of naïve $CD4^+$ T cells into Th2 cells *in vitro*. Moreover, treatment of peritoneal cell-derived MCs with INF-γ and IL-4 was shown to induce expression of MHC class II molecules on MCs (Gaudenzio et al. 2009). These MCs were able to present antigen to effector T cells causing their activation, proliferation, and formation of an immunological synapse between the MC and the T cell. Very recently, the antigen-presenting function has been shown to be restricted to a subset of three-week old pure BMMCs expressing both high levels of surface FcεRI and surface MHC class II (Gong et al. 2010). Collectively, studies indicate that MCs represent a rich source of co-stimulatory activity by expressing also molecules of the B7 family (ICOS-L, PD-L1, PD-L2, CD80, CD86), members of the TNF/TNF receptor families (OX40L, CD153, Fas, 4-1BB), CD28 and CD40 ligand (Nakae et al. 2006; Hershko and Rivera 2010). Evidence for a role of histamine in modulating T cell proliferation via H1 receptor stimulation is suggested by an *in vitro* study where co-cultures of MCs with helper T cells were seen to cause either increased or decreased T cell proliferation when low or high numbers of MCs, respectively, were challenged with helper T cells (Khan et al. 1986). MCs have recently demonstrated to regulate B cell survival and activation. Coculture assays using mouse splenic B cells and BMMCs revealed that both nonsensitized and activated MCs proved able to induce a significant inhibition of cell death and an increase in proliferation of naïve B cells (Merluzzi et al. 2010). Such proliferation was further enhanced in activated B cells. This effect relied on cell-cell contact and MC-derived IL-6. Activated MCs could regulate CD40 surface expression on unstimulated B cells, and the interaction between CD40 and CD40L on MCs, together with MC-derived cytokines, was involved in the differentiation of B cells into $CD138^+$ plasma cells and in selective IgA secretion. These data were corroborated by *in vivo* evidence of infiltrating MCs in close contact with IgA-expressing plasma cells within inflamed tissues.

MCs are key effector cells in initiating and/or amplifying IgE-dependent inflammatory reactions, including allergic disorders and certain protective immune responses to parasites (Gurish and Austen 2001; Galli et al. 2005a, 2008b). As a result, they have mainly been regarded in the past for their detrimental role in type I allergic reactions, such as anaphylaxis, hay fever, eczema or asthma. In addition, they also express immunoregulatory functions in the same settings (Gribaldeston et al. 2006; Galli et al. 2008a; Rauter et al. 2008). MCs are activated during IgE-associated anaphylaxis. Anaphylaxis is an acute-onset, potentially fatal systemic allergic reaction that can be triggered by immunological or non-immunological mechanisms (Estelle et al. 2008). IgE play a crucial role in the immediate hypersensitivity response through binding with the high-affinity receptor FcεRI but other IgE-independent mechanisms, such as G protein-coupled receptor and TLR activation processes may intervene (Marshall 2004; Vines and Prossnitz 2004). Activated

MCs, along with basophils, release Th2 cytokines (IL-4, IL-5, IL-9 and IL-13) that polarize the immune reaction, and produce various bioactive chemical mediators, such as histamine and lipid metabolites, that provide vasoactive, chemotactic and immunoregulatory functions. In addition to their roles in classic acute IgE-associated immediate hypersensitivity responses, several lines of evidence indicate that MCs can also contribute to late-phase and chronic allergic reactions (Holgate 2002; Galli et al. 2008b). MCs have been shown to change their degranulation pattern from acute to chronic allergic responses (Theoharides et al. 2007). Anaphylactic degranulation is triggered by IgE-dependent and neuropeptide activation mainly during the early phase of allergic reactions. Differential release without degranulation is activated by mediators such as IL-1 (in humans), SCF (in mice), LPS (in rats) and CRH (in humans), and classically occurs during chronic inflammatory diseases (Cao et al. 2005; Theoharides et al. 2007). Many clinical symptoms of IgE-dependent late-phase reactions, both in the respiratory tract, gastrostrointestinal tract and the skin, reflect the actions of the leukocytes recruited to these sites by MCs. Cytokines (TNF-α, IL-6, IL-8) and neutral proteases, as well as histamine and lipid mediators, may contribute to MC-dependent leukocyte recruitment—in particular eosinophil recruitment—in such settings (Puxeddu et al. 2005). Leukocytes, in turn, expand the inflammatory reaction by providing additional pro-inflammatory mediators and cytokines ("MC-leukocyte cytokine cascade"). MC cytokines, such as TNF-α, VEGF, FGF-2 and TGF-β, contribute to chronic allergic inflammation through effects on fibroblasts, vascular endothelial cells, and other cells resident at the sites of these reactions (Galli et al. 2008b). Persistent chronic allergic inflammation can result in remodelling of the affected tissues and these structural changes are often associated with activation of the angiogenic process.

Very recently, MCs have been associated with a new type of immune function that is the induction of immune tolerance. Participation of MCs to this kind of activity explains certain aspects of MC involvement in the dynamics of tumour development and progression. Although immune surveillance works at an early stage of tumorigenesis, the established tumours primarily induce immune tolerance, by creating sites of immune privilege and by inducing the shift of the immune balance from activation to tolerance (Pardoll 2003; Munn and Mellor 2006). In this perspective, MCs have recently been proposed to be mechanistically involved in the negative modulation of immune surveillance in the tumour microenvironment. MCs, indeed, can promote suppression of immune reactions not only by producing inhibitory cytokines, such as IL-10, but also by promoting the immune tolerance mediated by Treg cells (Ullrich et al. 2007; Grimbaldeston et al. 2007). Indeed, MCs serve as enforcers for Treg cells, turning down the immune system's reaction to skin allograft possibly by IL-10 secretion (Lu et al. 2006; Grimbaldeston et al. 2007). Ultraviolet B irradiation, which represents the most important skin immunosuppressor and initiator of cutaneous malignancies, activates MCs (Kripke 1984; Ch'ng et al. 2006). Upon irradiation of the skin, *trans*-urocanic acid in the epidermis isomerizes to *cis*-urocanic acid, which stimulates substance P and CGRP release from neural C-fibres. These neuropeptides, in turn, trigger secretion of histamine, TNF-α and other mediators from MCs, leading to suppression of the cellular immune system

(Wille et al. 1999; Hart et al. 2001, Khalil et al. 2001). Using the MC-deficient $W/W^v$ mice, a direct correlation was demonstrated between MC density in the dermis and susceptibility to ultraviolet-B-induced systemic immunosuppression (Hart et al. 2002). In a skin transplantation model of allograft tolerance in the mouse, MCs were crucial for graft acceptance as MC-deficient C57BL/6-$Kit^{W-sh/W-sh}$ mice showed inability to induce tolerance (Lu et al. 2006). Activated Treg cells in the tolerant tissue produced high levels of IL-9, a cytokine which seems important in MC recruitment, growth and activation. This cytokine appears a crucial factor in mediating regional tolerance because neutralization of IL-9 greatly accelerated allograft rejection in tolerant mice (Lu et al. 2006). In a mouse model of tumorigenesis, SCF-activated MCs exacerbated the immunosuppression in the tumour microenvironment (Huang et al. 2008). MCs were shown to promote the decrease of the immune activatory factor IL-2 mRNA in the tumour whereas inducing the increase of immune suppressory factors such as IL-10, TGF-β, and Foxp3 RNAs. A recent study suggests that MCs and Treg cells may cooperate with each other in hepatocellular carcinoma determining a poorer tumour evolution (Ju et al. 2009). Remarkably, the percentage of Treg cells was shown to increase because $CD4^+CD25^-$ T cells can be converted into $CD4^+CD25^+$ regulatory T cells by TGF-β expression of transcription factor Foxp3 (Chen et al. 2003). Indeed, in an ovalbumin peptide TCR transgenic adoptive transfer model, TGF-β-converted transgenic $CD4^+CD25^+$ suppressor cells proliferated in response to immunization and inhibited antigen-specific naive $CD4^+$ T cell expansion *in vivo*. In addition, in a murine asthma model, coadministration of these TGF-β-induced suppressor T cells has been shown to prevent house dust mite-induced allergic pathogenesis in lungs (Chen et al. 2003). Another mechanism of potential physiological and pathological significance in MC-driven Treg cell expansion might be linked to the LPS- and IFN-γ-induced expression of MHC class II on MCs. As previously reported, the expression of MHC class II grants MCs the ability to process and present antigens directly to T cells. Remarkably, MCs preferentially expand antigen-specific Treg cells over naïve T cells, a fact that may explain one of the mechanisms that governs allograft tolerance induction by MCs (Kambayashi et al. 2009). MCs, are poor stimulators of naïve T cells *in vitro*, however, the role for MHC class II expression on MCs may be to activate regulatory Treg cells and dampen the immune response or avoid self-reactivity. Activation of Treg cells by MCs may contribute to the protective effect of MCs on skin allografts (Lu et al. 2006; Kambayashi et al. 2009). Conversely, MCs are also able to counteract Treg-mediated suppression through IL-6 secretion and OX40/OX40L interaction (Piconese et al. 2009). This reversal Treg suppression leads to the establishment of Th17-mediated inflammatory responses. The cross-talk between Treg cells and MCs is bidirectional. Treg cells, indeed, have been shown to influence the immediate hypersensitivity response MCs. Treg cells directly inhibit the FcεRI-dependent MC degranulation through cell-cell contact involving OX40-OX40L interactions between Treg cells and MCs, respectively (Gri et al. 2008).

Finally, some words should be devoted to the role of MCs in autoimmunity. Several lines of evidence indicate that MC may play a role in autoimmunity, affecting disorders like arthritis, multiple sclerosis, bullous pemphigoid and Graves' ophthalmia. MCs help initiate rheumatoid arthritis (Lee et al. 2002). $W/W^v$ mice lacking

MC do not develop the rodent equivalent of this debilitating condition. In humans, an increased number of MCs are found in the synovial tissues and fluids of patients with rheumatoid arthritis and at the site of cartilage erosion, reflecting the presence of MC chemotactic or survival factors, such as SCF and TGF-β, in the synovial fluid (Olsson et al. 2001). The invading MCs show ultrastructural signs of cell degranulation and produce several inflammatory mediators, notably TNF-α, IL-1β and VEGF. TNF-α reportedly plays a pivotal role in the pathogenesis of rheumatoid arthritis, especially in its ability to regulate IL-1β expression, this being important for the induction of prostanoid and matrix metalloproteinase (MMP) production by synovial fibroblasts and chondrocytes. In addition, MCs promote leakage of fluids into the joints, which in turn allows penetration of self-targeted antibodies that might lead to tissue damage by activating the complement cascade (Nigrovic and Lee 2007). Growing evidence suggests that MCs play a crucial role in the inflammatory process and subsequent demyelinization observed in patients suffering from multiple sclerosis. Indeed, recent results from animal models with experimental autoimmune encephalomyelitis clearly indicate that these cells act at multiple levels to influence both the induction and the severity of the disease, possibly by enhancing Th1 cell response through secretion of IL-4 (Gregory et al. 2006; Christy and Brown 2007). Bullous pemphigoid is another human disease whereby MCs have been proposed to exert a relevant pathogenic role. This autoimmune skin disease is characterized by subepidermal blisters resulting from auto-antibodies against two hemidesmosomal antigens, BP230 and BP180. Intradermal injection of antibodies against BP180 into neonatal mice causes a blistering disease mimicking bullous pemphigoid. Injection of antibodies against BP180 into MC-lacking $W/W^{v}$ mice does not induce bullous pemphigoid, nor does the injection into wild-type mice pre-treated with the MC stabilizer cromolyn sodium induce it (Chen et al. 2002). Interstitial cystitis has gained increasing attention for an involvement of MC in its pathogenesis. Indeed, the presence of activated MC in close proximity to suburothelial nerves is a consistent feature of this yet-to-be-clarified urological pathology (Elbadawi 1997).

In conclusion, the picture sketched in this paragraph underlines the pivotal role played by MCs in different immunological contexts and experimental settings. These cells are increasingly being recognized as key elements in orchestrating complex defensive and immune reactions which usually protect the host but, sometimes, may turn out to be dangerous or even lethal. In this perspective, the importance of MCs as initiators and effectors of both innate and adaptive immunity in healthy individuals has recently been appreciated insofar as MC activation can be used as an adjuvant to promote Ag-specific humoral immune responses upon vaccination (Fang et al. 2010).

### *2.2.7 Mast Cell Non-immunological Functions*

MCs play an important role not only in immediate hypersensitivity and late phase inflammation but also in tissue remodelling in different organs. Evidence indeed has been accumulated that MCs may exert a series of relevant non-immunological

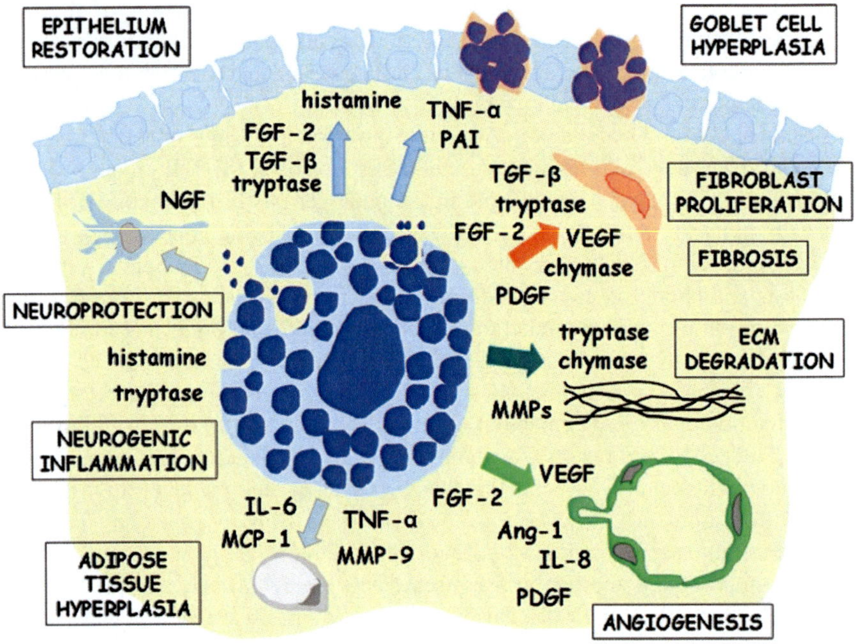


**Fig. 2.15** Mast cells couple in a strict temporal sequence inflammatory and healing processes such as tissue homeostasis, repair, remodelling, and fibrosis. Mast cell-derived cytokines, such as TNF-α, VEGF, FGF-2 and TGF-β, contribute to wound healing by acting on fibroblasts, vascular endothelial cells, and other cells resident at the sites of these reactions. Serine proteases, tryptase and chymase, as well as matrix metalloproteinases (MMPs) favour extracellular matrix degradation and remodelling. Mast cell mediators VEGF, FGF-2, Ang-1, IL-8 and PDGF, among others, stimulate angiogenesis. Neurogenic inflammation and neuroprotection are favoured by mast cell products like histamine, tryptase and NGF. Histamine, FGF-2, TGF-β and plasminogen activator inhibitor (PAI) stimulate epithelium restoration and goblet cell hyperplasia at the site of chronic inflammation. Remarkably, mast cell-derived products, such as IL-6, TNF-α, MMP-9 and MCP-1 play a crucial role in the induction of adipose tissue hyperplasia

functions, which couple in a strict temporal sequence inflammatory and healing processes, such as tissue homeostasis, repair, remodelling, and fibrosis (Fig. 2.15). The prototype of these *in vivo* experimental models is skin wound healing. Cutaneous wound healing is characterized by three sequential phases: (1) inflammation, as a direct consequence of wounding, (2) proliferation and (3) remodelling. Involvement of MCs in the various steps of cutaneous wound healing has long been recognized (Dvorak and Kissel 1991; Gruber et al. 1997; Metcalfe et al. 1997; Artuc et al. 1999, 2002). During the early phase of wound healing, MCs contribute to local coagulation, extravasation and leukocyte recruitment through secretion of histamine, TNF-α and other mediators. Proliferation of fibroblasts, endothelial cells and keratinocytes is sustained by MC products such as FGF-2, VEGF, TGF-β, histamine and tryptase. In particular, MCs can promote the conversion of fibroblasts into myofibroblasts, which facilitates wound contraction. The final step, tissue remodelling, is assisted by proteolytic, extracellular matrix-degrading enzymes such as MMPs produced by MCs. Formal demonstration of the essential role of MCs in the process of wound healing has been provided only recently using MC-deficient mice (Weller et al. 2006). In this study, experimentally induced skin wounds showed impaired closure in MC-deficient $W/W^v$ mice. In addition, $W/W^v$ mice showed diminished extravasation and recruitment of neutrophils to the wound areas. All these parameters were restored in MC-reconstituted $W/W^v$ mice. Remarkably, secretion of histamine, through H1 receptors, and TGF-α is essential for MC-mediated effective wound healing. Recent reports point to a protective role of MCs in murine anti-glomerular basement membrane glomerulonephritis (Hochegger et al. 2005; Kanamaru et al. 2006). These findings have been explained either in the light of the ability of MCs to engender repair mechanisms or by an immunomodulatory effect of MCs in the inflammatory setting. Although the interpretation of involved mechanisms may be different, the net result can be reconciled with the crucial role of MCs in maintaining tissue homeostasis. Conversely, MCs play a relevant role in airway remodelling and in the maintenance of airway hyperresponsiveness in asthmatic subjects. The extent of airway remodelling correlates with severity of asthma. The infiltration of the bronchial wall by MCs is associated with the disordered airway function. The increase in airway smooth muscle mass is recognized as one of the most important factors related to persistent airway hyperresponsiveness and to the severity of asthma. MC mediators such as tryptase, chymase, activin A, TNF-α, PDGF, TGF-β, plasminogen activator inhibitor (PAI) and amphiregulin can contribute to smooth muscle cell and goblet cell hyperplasia, thickening of the reticular basement membrane and angiogenesis (Okayama et al. 2007). MCs are found to contribute to the development of multiple features of chronic asthma in MC-deficient mice.

MCs have been closely linked with the development of fibrosis in several organ sites. Intestinal strictures associated with Crohn's disease have been shown to be densely populated by chymase-positive MCs. MC-derived chymase has been suggested to play a crucial role in fibrosis production by interfering with the local angiotensin II system (Suekane et al. 2010). In addition, chymase-expressing MCs are also prominent in scarring lung diseases, like interstitial lung disease (Edwards et al. 2005). Another mechanism of fibrosis induction may be mediated by the PDGF-osteopontin axis. Expression of osteopontin in wound fibroblasts is elicited by PDGF secreted by MCs and macrophages (Mori et al. 2008). Heparin and related glycosaminoglycans secreted by MCs differentially regulate PDGF-induced lung fibroblast proliferation, chemotaxis and MMP activity (Sasaki et al. 2000). These effects, in turn, may have a key role in extracellular matrix remodelling in inflammatory lung disease.

MCs may intervene to stimulate tissue growth and differentiation in health subjects. A role for MCs in promoting development of parenchymal tissues has been proposed since early Ehrlich's dissertation. In the last decade, MCs have increasingly been conceived as "feeding" and "nourishing" cells—a concept postulated by Paul Ehrlich—in a series of developmental settings. For instance, postnatal mammary gland branching morphogenesis in mice has been recognized to depend on MC functional integrity (Lilla and Werb 2010). Genetic and pharmacological disruption of MC function in the mammary gland reveals that MCs are involved in rapid proliferation and normal duct branching during puberty. Remarkably, MC degranulation and serine protease activation are required for normal mammary gland development. It is known that tissue-type MC proteases are implicated in angiogenesis and extracellular matrix protein degradation either directly or by activation of other proteases. The normal complement of active granule proteases is a critical factor in sustaining the growth of mammary gland ductal epithelium indicating that MCs contribute to the complex regulation of cell proliferation in the growing mammary gland. MCs are found throughout the mammary stroma as well as concentrated around the invading terminal end buds, a distribution pattern already recognized by Ehrlich in the developing goat parotid. Indeed, he wrote: "In certain acinar glands (goat parotid), the pattern of MC accumulation [inside the organ] is not determined by the branching of the vascular system but the ramification of the gland excretory ducts" (Ehrlich 1878). As normal development and tumorigenesis are closely linked processes, the potential role of MCs in facilitating mammary gland neoplasia has recently been investigated. It has been found that plasma kallikrein, a potent activator of the plasminogen cascade of serine proteases, is localized to CTMCs in the mouse mammary gland (Lilla et al. 2009). Active plasma kallikrein regulated ductal epithelial cell apoptosis, adipocyte differentiation and stromal remodelling during mammary gland involution. As plasminogen cascade directs tumorigenesis in the mammary gland, an active participation of local MCs in the neoplastic development may be suggested.

A contribution of MCs to diet-induced obesity and diabetes has recently been suggested. It has been shown that white adipose tissue from obese humans and mice contains more MCs than white adipose tissue from their lean counterparts (Liu et al. 2009). In addition, MC-deficient mice or mice treated with MC stabilizers show reduced body weight gain, reduced levels of inflammatory cytokines, chemokines and proteases, such as IL-6, TNF-α, IFN-γ, MCP-1, MMP-9 and cathepsin S in serum and white adipose tissue along with improved glucose homeostasis and energy expenditure. Thus, MCs are believed to contribute to diet-induced obesity and glucose intolerance by promoting adipose tissue protease expression and angiogenesis, which may favour adipose tissue growth.

MC involvement in the pathogenesis of coronary spasm, cardiomyopathy, atherosclerosis and myocardial ischemia has recently been suggested. It has been shown that chymase cleaves angiotensin I to angiotensin II more effectively than the angiotensin-converting enzyme (Church and Levi-Schaffer 1997). Studies in the canine model of myocardial ischemia and reperfusion indicate a role for MC mediators in initiating the cytokine cascade which is ultimately responsible for neutrophil accumulation in the ischemic area. In addition, MCs have been claimed to play a crucial role for leading to the subsequent fibrotic process (Frangogiannis et al. 1998). By using C57BL/6-$Kit^{W\text{-}sh/W\text{-}sh}$ mice crossed with atherosclerosis-prone mice deficient in low-density lipoprotein receptor, *in vivo* evidence has been obtained that MCs is implicated in the atherogenic process, as MC absence causes smaller atherosclerotic lesions with fewer inflammatory cell infiltrates (Sun et al. 2007a).

MC may also contribute to the pathogenesis of elastase-induced abdominal aortic aneurysms in mice, as C57BL/6-$Kit^{W\text{-}sh/W\text{-}sh}$ mice fail to develop such aneurysms (Sun et al. 2007b).

It has been reported that MCs intervene in the mechanisms controlling intestinal permeability. In particular, it has recently been discovered that CRH acts on intestinal MCs causing an increase of mucosal permeability to horseradish peroxidase (Wallon et al. 2008). The increased permeability to horseradish peroxidase was abolished by the MC stabilizer, lodoxamide. In this study, electron microscopy showed transcellular passage of horseradish peroxidase through colonocytes. CRH receptor subtypes R1 and R2 were detected in the HMC-1 cell line and in lamina propria MCs in human colon, suggesting that CRH mediates transcellular uptake of horseradish peroxidase in human colonic mucosa via CRH receptor subtypes R1 and R2 on subepithelial MCs. CRH-induced macromolecular uptake in human colon mucosa may have implications for stress-related intestinal disorders.

MCs have long been identified in the central nervous system, whereby they have been implicated in the regulation of permeability of the blood-brain-barrier (Esposito et al. 2002). In addition, it has been suggested that MCs in the central nervous system may function as a gate to the hypothalamic-pituitary-adrenal axis, thus participating in the counter-regulation of inflammatory immune responses (Matsumoto et al. 2001). Blood-brain-barrier permeability and multiple sclerosis appear to worsen in response to acute stress that leads to the local release of CRH, which activates brain MCs to selectively release IL-6, IL-8 and VEGF (Theoharides and Konstantinidou 2007). Recently, participation of MCs to blood vessel growth and differentiation in the rat nervous system during postnatal development has been proposed (Khalil et al. 2007). Here, MCs occur in two locations, namely the pia mater and the brain parenchyma. MC population in the pia reaches a maximum at postnatal day 11, and declines rapidly thereafter. In contrast, the thalamic MC population expands from postnatal day 8 to reach adult levels at postnatal day 30. Stereological analysis demonstrates that MCs home to blood vessels. Indeed, more than 96% of MCs are inside the blood-brain barrier, with ~90% contacting the blood vessel wall or its extracellular matrix. MCs express α4 integrins which represents a potential mechanism for adhesion to the vascular wall. At all ages studied, MCs are preferentially located on large diameter vessels (>16 μm; possibly arteries), and contact only those maturing blood vessels that are ensheathed by astroglial processes. MCs not only home to large vessels but also maintain a preferential position at branch points, sites of vessel growth. This observation presents the possibility that MCs participate in and/or regulate vasculature growth or differentiation.

Numerous microanatomical and ultrastructural investigations over the past years have demonstrated the presence of close nerve-MC contacts in different organs (skin, intestine, peritoneum) of various species, including man (Stead et al. 1987, 1989; Crivellato et al. 1991). It has also been reported that electrical stimulation of nerve fibers causes degranulation of tissue MCs and that these effects are inhibited by atropine or prior treatment with capsaicin (Javed et al. 1992). For instance, electrical vagal stimulation was observed to induce gastric mucosal MC degranulation. These findings have prompted further studies on a possible functional interaction between the peripheral nerve system and the tissue MC populations. It has subsequently been shown that several neurotransmitters and neuropeptides are able to modulate MC function and to induce granule release (Foreman and Jordan 1980). MCs in turn are able to stimulate nerve fibers through histamine release, thus amplifying the nerve-MC loop, or conversely decrease the local effects of nerve mediators by releasing neuropeptide-degrading proteases such as tryptase and chymase. In the skin, MCs have been found in close anatomical and functional association with sensory nerves, a condition which may explain neurogenic inflammation. During psychological stress the neuroendocrine system and peripheral sensory nerves are activated leading to release of mediators, such as neuropeptides, neurotrophins, CRH and α-melanocyte-stimulating hormone, which are capable of activating MCs (Harvima et al. 2010). On the other hand, MC mediators released, e.g. histamine, tryptase and NGF, can in turn excite and stimulate surrounding neuropeptide-containing C-fibers possibly resulting in feed-forward loop and potentiation of neurogenic inflammation. In these mechanisms, proinflammatory cytokines and chemokines are released from MCs. In chronic skin diseases, such as psoriasis, atopic dermatitis and palmoplantar pustulosis, the contacts between tryptase-positive MCs and sensory nerves are increased in number, which provides the morphological basis for increased MC-sensory nerve interaction in chronically inflamed skin. A key link between the neural tissue and MCs is NGF. The first evidence that MCs are receptive to NGF showed that exogenous administration of NGF in newborn rats induced a marked increase in the number and size of MCs in peripheral tissues (Aloe and Levi-Montalcini 1977). The effect of NGF on MCs includes stimulation of proliferation, differentiation, survival and mediator secretion. Furthermore, proliferation and differentiation of murine CTMCs has been shown to be dependent on the MC degranulation property of NGF (Matsuda et al. 1991). It has been ascertained that MCs, in turn, synthesize and release NGF (Nilsson et al. 1997). NGF appears to be increased in the circulation in a variety of inflammatory and autoimmune conditions; it most consistently appears to be elevated in the circulation of patients with multiple allergic diseases, including asthma (Bonini et al. 1996).

Very recently, a role of MCs in the bone metabolism has been suggested. In particular, MCs seem to be mechanistically involved in the formation of skeletal defects observable during chronic hyperparathyroidism, which is a common cause of metabolic bone disease. Bone biopsies from hyperparathyroid patients revealed an association between parathyroid bone disease and increased numbers of bone marrow MCs. In addition, mature MCs in rats were preferentially located at sites undergoing bone turnover, and the number of MCs at the bone-bone marrow interface was greatly increased following treatment with parathyroid hormone (Turner et al. 2010). Time-course studies revealed that mature MC redistribution from bone marrow to bone surfaces precedes and is associated with osteitis fibrosa, a hallmark of parathyroid bone disease. Importantly, mature MCs were not observed in the bone marrow of mice, a species that is resistant to the development of parathyroid hormone-induced bone marrow fibrosis. These findings suggest that the MCs may play a crucial role in metabolic bone disease.

# Chapter 3
# Mast Cell Mediators

MCs are tissue-based immune cells that respond to signals of innate and adaptive immunity as well as to non-immunological stimuli. The outcome of MC stimulation by multiple triggers is the immediate and delayed release of biologically active mediators. Mediators produced by MCs are conventionally divided into (1) preformed mediators, (2) newly synthesized lipid mediators, and (3) cytokines/chemokines. Indeed, when MCs are activated they immediately extrude granule-associated substances, such as histamine and, within minutes, generate lipid-derived mediators (Galli et al. 2005a). Within hours, MC activation is followed by the *de novo* synthesis of numerous cytokines and chemokines. These categories are not absolutely exclusive because at least one cytokine, TNF-α, occurs both preformed and as a newly synthesized molecule.

## 3.1 Granule Composition

Secretory granules of MCs contain crystalline complexes of preformed mediators ionically bound to a matrix of proteoglycans. The metachromatic staining of MC granules probably reflects their content of proteoglycans, such as chondroitin sulphates and heparin. In the mouse, the proteoglycan content of MC granules varies in the different MC subtypes. CTMCs contain heparin that lacks in MMCs. Conversely, MMCs express chondroitin sulphates A and B, which are not found in CTMCs, whereas both MC subtypes store chondroitin sulphate E in their granules (Féger et al. 2002). The dominant proteoglycan in human MCs is heparin, which constitutes some 75% of the total, with a mixture of chondroitin sulfates making up the remainder (Church and Levi-Schaffer 1997). In humans, the heparin content in $MCs_T$ and $MCs_{TC}$ is roughly the same. Chondroitin sulphate and heparin proteoglycans are thought to bind histamine, neutral proteases, and carboxypeptidases primarily by ionic interactions and, therefore, contribute to the packaging and storage of these molecules in the granules. Mice that lack the enzyme N-deacetylase/N-sulphotransferase-2 (NDST-2), which are unable to product fully sulphated heparin, exhibit severe defects in the granule structure of MCs, with impaired storage of cer-

D. Ribatti, E. Crivellato, *Mast Cells and Tumours*,
DOI 10.1007/978-94-007-1469-4_3, © Springer Science+Business Media B.V. 2011

tain proteases and reduced content of histamine (Humphries et al. 1999; Forsberg et al. 1999). During degranulation, the various mediators packaged with proteoglycans dissociate at different rates, histamine very rapidly but tryptases and chymases much more slowly.

## 3.2 Histamine and Serotonin

Histamine, the first discovered mediator in MCs, is present at a concentration of 1–4 pg/cell in human MCs. Histamine exerts many effects pertinent to the immediate phase of allergic response, including vasodilation, increased vasopermeability, contraction of bronchial and intestinal smooth muscle cells, increased mucous production and stimulation of nerve endings (Fig. 3.1). Histamine mediates its action via the histamine receptors 1–4 (H1–4 receptors), which can transduce signals leading to the variety of symptoms associated with acute allergic reactions (Galli et al. 2008a). Upon MC activation, secretory granules fuse with the plasma membrane and release their content in the extracellular fluid within minutes. Histamine is rapidly liberated in the extracellular environment following dissociation from the proteoglycans of the granule matrix by sodium salt exchange. Histamine has a half-life of around one minute in the extracellular fluid and is degraded by the enzymes histamine N-methyltransferase and by diamine oxidase. In addition to its well-characterized effects in the acute inflammatory and allergic responses, histamine has been shown to affect chronic inflammation and regulate several essential events in the immune response. Histamine can selectively recruit the major effector cells into tissue sites and affect their maturation, activation, polarization, and effector functions leading to chronic inflammation through engagement of H4R, H1R and H2R (Jutel et al. 2009). Besides vasoactive and bronchoconstrictor effects mediated by H1R, histamine secreted by MCs has been shown to exert an important role in modulating T cell proliferation via H1R and H4R stimulation (Khan et al. 1986). Histamine also affects epithelial cells, dendritic cells and B lymphocytes. Histamine promotes Th1 cell activation through H1R and suppresses both Th1 and Th2 cell activation through H2R (Cherwinski et al. 2005). These findings provide suitable explanation for the observations in the experimental model of asthma showing that allergic inflammatory responses and bronchial hyperresponsiveness may be susceptible to H1R blockade. Secretion of histamine, by acting through H1R, has recently been shown to be essential for MC-mediated wound healing in a MC-deficient mouse model (Weller et al. 2006). These data support reports in the older literature that topical administration of histamine can accelerate wound healing (Dabrowski et al. 1975). During the process of wounding, MC granules released into the tissue are phagocytozed by fibroblasts and endothelial cells and might thus contribute to persistently increased tissue histamine levels (Seibold et al. 1990). Histamine has an angiogenic effect through both H1R and H2R (Sorbo et al. 1994). It may also increase the permeability of newly formed microvessels during tumour angiogenesis, and hence increase the leakage of plasma proteins and deposition of

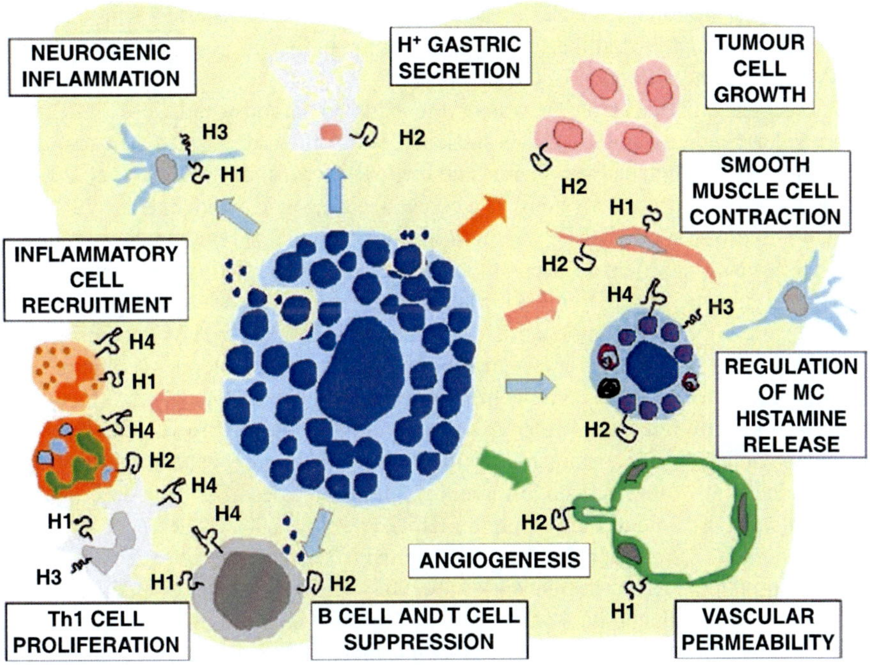


**Fig. 3.1** Schematic drawing which illustrates the effects that histamine liberated by mast cells exerts on different tissue structures and cell types. During type I allergic reactions, such as anaphylaxis, hay fever, eczema or asthma, histamine promotes capillary permeability and smooth muscle cell contraction through the H1 receptor. In IgE-dependent late-phase reactions, both in the respiratory tract, gastrostrointestinal tract and the skin, histamine stimulates angiogenesis and basophil chemotaxis through the H2 receptor. This biogenic amine has relevant function on inflammatory cell recruitment through the H1–H4 receptors. In particular, the H4 receptor exerts T, B and dendritic cell activation. Conversely, the H2 receptor effects suppression of T and B cell functions. The H1 and H3 receptors are involved in neurogenic inflammation, whilst H2 receptor mediates $H^+$ gastric secretion. Remarkably, histamine activates tumour cell growth through H2 receptor which in addition, promotes immune suppression. Histamine also inhibits or potentiates mast cell activation and histamine liberation through H2 and H4 receptors, respectively, whilst the H3 receptor is involved in mast cell activation by nerve endings

fibrin. In addition, histamine has recently been recognized to promote skin cancer in a cutaneous model of cancer. In mice, dermal MCs are critical to UVB-induced systemic immunomodulation. In this species, a functional link as well as a linear relationship has been identified between the prevalence of histamine-staining dermal MCs and the log of the dose of UVB required for 50% immunosuppression (Hart et al. 2001). Studies with histamine receptor antagonists support histamine as the main product of MC involved. Histamine may have multiple roles, but experiments with indomethacin administered to mice have shown that one process involves induction of prostanoid production (Hart et al. 2000). As previously referred, MCs in the central nervous system have been suggested to function as a gate to the hypothalamic-pituitary-adrenal axis, thus participating in the counter-regulation of inflammatory immune responses (Matsumoto et al. 2001). This effect would be accomplished through histamine secretion, activation of histamine H1 receptors at the hypothalamus and induction of CRH.

The biogenic amine serotonin was first recognized in mouse and rat MC granules (Bergendorff and Uvnäs 1973). This substance was further localized to MC granules of guinea-pig and man as well (Lehtosalo et al. 1984; Kushnir-Sukhov et al. 2007). Serotonin is produced in two steps. The essential amino acid tryptophan is hydroxylated to 5-hydroxytryptophan by the rate-limiting enzyme that regulates the synthesis of 5-HT, tryptophan hydroxylase, which has been demonstrated in both normal and neoplastic human MCs (Slominski et al. 2003; Kushnir-Sukhov et al. 2007). In a second step, 5-hydroxytryptophan is decarboxylated to form 5-HT. Metabolism by monoamine oxidase is the primary metabolic pathway for serotonin. Serotonin and histamine are stored in the same compartment as serglycin proteoglycan (Ringvall et al. 2008). Serotonin is implicated in immunomodulation, cell growth and development, MC adhesion and chemotaxis, tumorigenesis and tissue regeneration (Mohammad-Zadeh et al. 2008). Fifteen different receptor molecules have been described in mammals that are able to bind serotonin (Bockaert et al. 2006). The best characterized is the 5-$HT_{1A}$ receptor that has been shown to mediate the effects of 5-HT in murine and human MCs (Lundeberg et al. 2002; Kushnir-Sukhov et al. 2007). Expression of serotonin and its 5-$HT_{1A}$ receptor has been reported in canine cutaneous MC tumours (Fröberg et al. 2009). Serotonin is reported to act as a growth factor for several types of human cancer (Vicaut et al. 2000; Siddiqui et al. 2005). Serotonin can produce itching, flare and wheals when applied to human skin. Serotonin induces bronchoconstriction (Metcalfe 2008) and also acts as a chemoattractant for eosinphils (Boehme et al. 2004). Indeed, the 5-$HT_{2A/2C}$ receptor antagonist ketanserin fully prevents the compound 48/80-induced bronchoconstriction in rats (Krop et al. 2010). Recently a relation between serotonin in MCs and antizyme inhibitor 2 (AZIN2), an activator of polyamine biosynthesis, has been recognized both in resting and activated MCs (Kanerva et al. 2009). Immunostainings show that AZIN2 is expressed in primary and neoplastic human and rodent MCs, and localizes in the Vamp-8 positive, serotonin-containing subset of MC granules, but not in tryptase-containing granules, as revealed by double immunofluorescence stainings. Furthermore, activation of MCs induces rapid upregulation of AZIN2 expression and its redistribution, suggesting a role for AZIN2 in secretory granule exocytosis. In addition, the release of serotonin from activated MCs is polyamine-dependent whereas release of histamine and β-hexosaminidase is not, indicating a granule subtype-specific function for polyamines as selective regulators of serotonin release from MCs.

## 3.3 Proteases

MCs are a rich reservoir of neutral proteases. They are packed in large amounts in the granules and comprise a high fraction of all cellular proteins. The principal proteases are tryptases, chymases, cathepsin G, carboxypeptidase A3 and dipeptidylpeptidase I (also known as cathepsin C) (for a review of this issue, see Triverdi and Caughey 2010). Recent studies indicate that these enzymes are crucial for host defence and homeostasis. Indeed, they are essential for innate antibacterial inflammatory responses and for tissue remodelling after injury. MC protease expression is modulated by cytokines, such as SCF, TGF-β, IL-10 and IL-3 (Gurish et al. 1992; Ghildyal et al. 1992; Miller et al. 1999).

The major class of MC proteases are the tryptases, a group of 130 kD serine peptidases, which are stored in a fully active form in the granule. They represent the most abundant constituent of human MCs. Some 10 pg/cell has been detected in MCs in the lung and up to 35 pg/cell in skin MCs (Schwartz et al. 1987a). Tryptases are clinically used as markers of mastocytosis and systemic MC activation, such as anaphylaxis (Schwartz et al. 1987b). This class of proteases play an important role in extracellular matrix degradation, contributing to processes like wound healing and tumour metastasis. There are numerous endogenous targets for tryptases. Indeed, they cleave: (1) extracellular matrix components such as fibronectin and type VI collagen, (2) pre-enzyme forms of some MMPs and uPA, (3) various bronchial and intestinal neuropeptides such as CGRP and vasoactive intestinal peptide (VIP), and (4) IgE molecules thus possibly down regulating the allergic response (Rauter et al. 2008). There are other potential endogenous targets of tryptases, although the most important targets may be exogenous and pathogen-associated (Triverdi and Caughey 2010). In addition to this, tryptases express important functions as potent growth factors for fibroblasts, endothelial cells and muscle cells (Blair et al. 1997; Gruber et al. 1997). Tryptases have been shown to be potent activators of fibroblast migration and proliferation (Ruoss et al. 1991; Artuc et al. 2002) stimulating tissue repair in wound healing and fibrosis. They can induce the synthesis and release of collagen from fibroblasts in culture, as well as provoking secretion of collagenase (Cairns and Walls 1997). They can also induce the proliferation of airway smooth muscle cells, contributing to the condition of smooth-muscle cell hyperplasia in bronchial asthma (Thabrew et al. 1996). Tryptases have a potent proangiogenic potential, stimulating endothelial cell activation, proliferation, migration and tube formation (Blair et al. 1997). Interestingly, MC tryptase has recently been demonstrated to degrade the antiangiogenic molecule endostatin *in vitro* (Syväranta et al. 2010). In the mouse, at least three different granule-associated tryptases (mMCP-6, mMCP-7, mMMP-11/transmembrane tryptase [mTMT]) have been described at the protein level (Huang et al. 1998). There appear to be multiple forms of human tryptases as well (tryptases α, I, II/β, III) (Vanderslice et al. 1990; Miller et al. 1989, 1990). The α- and β-tryptases have 90% sequence homology. Mature β-tryptase is the predominant form stored in secretory granules of all human MCs. It consists of four monomers stabilized in the tetrameric form by heparin proteoglycan. Recently, MC tryptases have been shown to play important roles in host defence, being a critical link between adaptive and innate immunity. MC tryptase mMCP-6, for instance, has a critical protective function in bacterial and parasite infection. MC-deficient mice pretreated with human tryptase defend themselves more effectively against intratracheally delivered *Klebsiella pneumoniae* (Huang et al. 2001). mMCP-6-deficient mice are less able to clear *Klebsiella pneumoniae* injected into their peritoneal cavities, probably because of less recruitment of neutrophils (Thakurdas et al. 2007). Delayed expulsion of the adult helminth and increased deposition of larvae in muscles occur in mMCP-1-deficient mice infected with *Trichinella spiralis* (Knight et al. 2000). mMCP-6 as well is important for the clearance of the chronic *Trichinella spiralis* infection (Shin et al. 2008).

Chymases, a group of 30 kD proteases, are present within the granules of the $MC_{TC}$ subset of MCs, in an estimated concentration of 4.5 pg/cell (Schwartz et al. 1987b; Metcalfe et al. 1997). Like tryptases, chymases are capable to target both endogenous and exogenous targets. They degrade some neuropeptides and interleukins such as IL-6 and IL-13 (Zhao et al. 2005). They cleave type IV collagen and other extracellular matrix components, being capable to split the dermal-epidermal junction (Huang et al. 1998). They activate tissue MMPs like MMP-9 and MMP-2 (Tchougounova et al. 2005). Chymases cleave peptides like angiotensin I and hepatocyte growth factor, and blood proteins like albumin (Reilly et al. 1982; Raymond et al. 2003). They also cleave exogenous molecules like the allergen profilin (Mellon et al. 2002). Chymases may also express a proangiogenic activity, thus contributing to tissue remodelling. In the mouse, at least five different granule-associated chymases (mMCP-1, mMCP-2, mMCP-3, mMCP-4, mMCP-5) have been described at the protein level (Huang et al. 1998). As we have previously mentioned, recent evidence indicates that mMCP-2, can contribute to neutrophil recruitment and host survival during cecal ligation and puncture in mice (Orinska et al. 2007).

Two other proteases, cathepsin G and carboxypeptidase A3, have been associated with the $MC_{TC}$ subset of human MCs. Whilst cathepsin G is expressed in a variety of leukocytes, including neutrophils, monocytes and dendritic cells, carboxypeptidase A3 appears to be largely specific of MCs (Irani et al. 1991). Carboxypeptidase A has been possibly linked to snake venom degradation. A recent study in mice has found that MCs can significantly reduce snake venom-induced pathology (Metz et al. 2006). Interestingly, the specific protease content of individual MCs can vary depending on the MC microenvironment. For example, MMCs in mice express mMCP-1 and mMCP-2, whereas CTMCs express a different pattern of proteases, namely mMCP-3, mMCP-4, mMCP-5, mMCP-6, mMCP-7 and carboxypeptidase (Stevens et al. 1993; Miller and Pemberton 2002). Degradation of the extracellular matrix occurs under physiological and pathological conditions. This process is principally mediated by a family of neutral proteolytic enzymes termed the MMPs. MCs synthetize and release different MMPs. Both mouse and human MCs are capable of producing MMP-9, a matrix-degrading enzyme necessary for leukocyte transmigration. Bacterial LPS enhances MMP-9 production of mouse BMMCs (Tanaka et al. 2001). In addition, IgE crosslinkage of FcεRI induces both production and activation of MMP-9 in BMMCs (Tanaka and Matsuda 2004). MMP-2, MMP-9 and MMP-13 can degrade type IV collagen which is the major component of the basement membrane zone (BMZ). In bullous pemphigoid, the separation occurs within the BMZ (Niimi et al. 2006). Interestingly, MCs are also able to synthetize tissue inhibitors of MMPs (TIMPs). Koskivirta et al. (2006) using western blotting and immunocytochemistry, were able to demonstrate that human blood derived MCs produce TIMP-4, which may play an important role in human cardiovascular disorders. Murine BMMCs secrete TIMP-1 (Frank et al. 2001). MC chymase cleaves and activates progelatinase B. Outside of cells, progel B is complexed with TIMP-1, which hinders zymogen activation and inhibits activity of mature forms. Interestingly, Latexin, a carboxypeptidase A inhibitor, is expressed in rat peritoneal MCs and is associated with granular structures distinct from secretory granules and lysosomes (Uratani et al. 2000). Very recently, it has been demonstrated that cyclic adenosine monophosphate induces plasminogen activator inhibitor-1 expression in the human MC line HMC-1 (Ma et al. 2010).

## 3.4 Lipid Mediators

The most important MC-derived lipid mediators are cyclooxygenase and lipoxygenase metabolites of arachidonic acid (Galli et al. 2005a). Upon FcεRI or KIT activation, MCs rapidly synthesize eicosanoid mediators from endogenous membrane arachidonic acid stores and lipid bodies. All these products have potent inflammatory activity and can also modulate the release process. The major cyclooxygenase product of MCs is $PGD_2$ which is not produced in basophils. $PGD_2$ is a bronchoconstrictor and attracts eosinophils and basophils. Its active metabolite, 9α, 11β-$PGF_2$ is a constrictor of coronary arteries. $PGD_2$ released by MCs during allergic reactions can also regulate T cell function. At the onset of asthmatic attack, $PGD_2$ causes airway hypersensitivity and chemotaxis of T cells, basophils and eosinophils through interaction of two receptors, the prostanoic DP receptor (PTGDR) expressed on granulocytes and smooth muscle cells, and CRTH2 (chemoattractant receptor-homologous molecule expressed on Th2 cells) on Th2 cells (Brightling et al. 2002; Oguma et al. 2004). The major lipooxygenase products derived from MCs are the LTs: $LTC_4$ and its peptidolytic derivates $LTD_4$ and $LTE_3$. Human MCs can also product $LTB_4$, although in much smaller quantities than $PGD_2$ or $LCT_4$, and some MC populations represent a source of PAF. MC-derived $LTB_4$ is essential for the recruitment of both $CD4^+$ and $CD8^+$ effector cells as well as for chemotaxis of neutrophils to sites of inflammation (Ott et al. 2003). $LTC_4$, $LTD_4$ and $LTE_4$ are potent bronchoconstrictors, promote vascular permeability, induce mucous production, and attract eosinophils. PAF, released from MCs and basophils during allergic reactions, causes bronchoconstriction and vascular permeability. Patients with severe anaphylaxis might have higher levels of serum PAF. MC "heterogeneity" is recognizable also in the pattern of secretion of lipid mediators. For instance, human synovial MCs obtained from osteoarthritis specimens are capable of elaborating approximately equivalent amounts of $PGD_2$ and $LTC_4$ when stimulated by IgE cross-linking, unlike MCs in the skin that synthesize $PGD_2$ preferentially (de Paulis et al. 1996). A PAF-mediated amplification loop for MC activation in the generation of anaphylaxis has recently been described (Kajiwara et al. 2010). Indeed, PAF induces histamine release from human lung MCs and peripheral blood-derived MCs but not skin MCs. Activation of PAF receptor-coupled $G_{\alpha i}$ leads to degranulation through phospholipase C-γ1 and phospholipase C-β2 activation in human MCs.

PAF-induced degranulation is rapid, being maximal at 5 seconds, and is partially dependent on extracellular $Ca^{2+}$.

## 3.5 Cytokines and Growth Factors

Human and mouse MCs synthesize and release an array of cytokines and growth factors involved in inflammation, immunity, haematopoiesis, tissue remodelling and other biological functions. More than thirty different cytokine mediators have been shown to be produced by human and mouse MCs. Human and mouse MCs secrete TNF-α, TGF-β, FGF-2, VEGF, GM-CSF, NGF, PDGF, IFN-α, -β and -γ, many interleukins such as IL-1α, IL-1β, IL-3, IL-4, IL-5, IL-6, IL-8, IL-9, IL-10, IL-11, IL-12, IL-13, IL-15, IL-16, IL-18, IL-25. MC secretory granules contain pools of stored TNF-α which has pleiotropic proinflammatory effects (Gordon and Galli 1990, 1991). TNF-α has been implicated in neutrophil recruitment, inducing up-regulation of the endothelial-leukocyte adhesion molecule (ELAM-1) (Walsh et al. 1991). TNF-α has also been known to enhance the bactericidal activities of neutrophils (Kenny et al. 1993). In addition, human MCs have the capacity to generate IL-8, thus contributing to neutrophil recruitment (Moller et al. 1993). Bacterial and fungal pathogens are able to induce a highly selective production of IL-1β, GM-CSF and LTs from human MCs through TLR receptor stimulation (McCurdy et al. 2003). It has been shown that human MCs can be activated by different HIV-1 proteins (gp120 and Tat) and thereby represent a potentially important source of Th2 cytokines during HIV-1 infection (Marone et al. 2000). Under allergic conditions MCs produce significant amounts of IL-1 that may contribute to lymphatic infiltration (Bochner et al. 1990) and IL-4, essential for the triggering of Th2 lymphocytes that themselves produce IL-4 to initiate inflammatory cell accumulation and B lymphocyte immunoglobulin class switching to IgE (Bradding et al. 1993). Other MC cytokines found in normal and in asthmatic airways are IL-5 and IL-6 which, together with IL-4 and IL-13, would enhance Th2-type immune response and eosinophil chemotaxis, thus indicating that MCs may play an important role in initiating and maintaining the inflammatory response in asthma (Bradding et al. 1995). Interestingly, a unique profile of cytokines is induced depending upon the nature of the stimulus or type of infection. Human intestinal MCs have been shown to spontaneously produce proinflammatory cytokines such as TNF-α, IL-6 and IL-8 at low levels without stimulation of the cell (Lorentz and Bischoff 2001). Stimulation by IgE receptor cross-linking leads to an enhanced production of proinflammatory cytokines and *de novo* production of Th2 cytokines such as IL-3, IL-5, IL-10 and IL-13. Gram-negative bacteria, in contrast to IgE receptor cross-linking, do not induce the release of Th2 cytokines but enhance that of proinflammatory cytokines. Remarkably, mouse MCs store and release IL-15 that negatively regulates the bactericidal activity of mouse MC chymase (mMCP-2) (Orinska et al. 2007). Interestingly, MCs from different anatomical sites are able to generate distinct profiles of cytokine expression. In the tissues of bronchial and nasal mucosae from normal, asthmatic and allergic rhinitis patients, $MCs_T$ release IL-5, IL-6 and some IL-4, whereas $MCs_{TC}$ preferentially express IL-4 but little IL-5 and IL-6. A similar predominant IL-4 pattern is recognizable in skin MCs which contain both tryptase and chymase ($MCs_{TC}$). Such differences in the distribution of cytokine expression between subsets of MCs suggest a difference in the capacity of MC subsets to produce various cytokines and therefore a difference in their specific roles in allergic inflammation.

The ubiquitous presence of MCs in connective tissues has been linked to the mechanisms of tissue repair after injury through the production of cytokines and growth factors, such as VEGF, FGF-2, TNF-α, TGF-β and PDGF. The spectrum of cytokines expressed appears to vary depending on the maturity state of the MCs and of the tissue of residence. MCs synthesize and release VEGF, the most potent proangiogenic mediator known so far (Grutzkau et al. 1998). VEGF-A, VEGF-B and VEGF-C have been recognized in all mouse and human MCs examined so far (Detoraki et al. 2009). Certain MC populations release preformed stores of VEGF after IgE-dependent upregulation of FcεRI expression (Boesiger et al. 1998). Murine MCs lacking JunB, a member of the AP-1 transcription factor family regulating IgE-mediated MC degranulation, display a severe impairment of *in vitro* angiogenesis due to inhibition of VEGF secretion (Textor et al. 2007). Human MCs, however, are a potent source of VEGF in the absence of degranulation through activation of the EP(2) receptor by $PGE_2$ (Abdel-Majid et al. 2004) Selective release of VEGF by human MCs is regulated by CRH (Cao et al. 2006). FGF-2 is a potent activator of fibroblast migration and proliferation (Artuc et al. 2002). It has been shown to localize to cytoplasmic and extruded granules of MCs in several human tissues (Qu et al. 1998). In an *in vivo* model of wound healing, an increased number of MCs has been detected to positively stain for FGF-2 during the fibroproliferative stage (Liebler et al. 1997). FGF-2 is also a potent proangiogenic factor. In the context of remodelling processes, it has been shown that TNF-α derived from MCs can promote elongation of cutaneous nerves during contact hypersensitivity in mice and, in addition, stimulates angiogenesis (Kakurai et al. 2006). TGF-β exerts a variety of effects on wound repair including the induction and/or facilitation of directed cell migration, stimulation of myofibroblast proliferation, angiogenesis and granulation tissue formation. In addition, TGF-β exerts a potent chemotactic effect on MCs (Gruber et al. 1994). MCs are capable of both responding to and producing TGF-β. Moreover, latent TGF-β bound to extracellular matrix can be released but not activated by MC-derived chymase (Taipale et al. 1995). MCs release PDGF into wounded tissue and thereby influence the healing process from very early stages onward. In addition, MCs synthesize and release NGF, which actively stimulates neurogenesis after injury. Indeed, in the rat intestinal mucosa, reconstitution of nerve fibres after experimentally-induced inflammation and nerve fibre degeneration is accompanied by a significant increase in mucosal MC density (Stead et al. 1991). NGF also induces endothelial cell proliferation *in vitro* and angiogenesis *in vivo* in the chick embryo chorioallantoic membrane (CAM) assay (Cantarella et al. 2002; Emanueli et al. 2002). NGF appears to be increased in the circulation in a variety of inflammatory and autoimmune conditions; it most consistently appears to be elevated in the circulation of patients with multiple allergic diseases, including asthma (Bonini et al. 1996). Recently, MCs from human uterine leiomyomas have been found to contain leptin, a 167-amino-acid residue peptide mainly secreted by adipocytes which, besides its involvement in obesity development, expresses angiogenic activity (Ribatti et al. 2007a).

Nakayama et al. (2004) reported that primary murine MCs express Ang-1 but not Ang-2. An *in vivo* angiogenesis assay using extracellular matrix components showed that MCs and plasmacytoma cells, together, promote marked neovascularization composed of dilated vessels, which was prevented by neutralization of VEGF-A and Ang-1 but was only partially reduced by neutralization of either VEGF-A or Ang-1. MCs within extracellular matrix components expressed Ang-1, and recombinant Ang-1 together with plasmacytoma cells promoted extracellular matrix neovascularization similar to that induced by MCs. A transplantation assay shows that primary MCs accelerate tumour growth by established plasmacytoma cell lines and that neutralization of Ang-1 alone or with VEGF-A reduces significantly the growth of plasmacytomas containing MCs. These results demonstrate that MC-derived Ang-1 promotes the growth of plasmacytomas by stimulating neovascularization and provide further evidence supporting a causal relationship between inflammation and tumour growth.

Adrenomedullin, a 52-amino acid peptide originally isolated from human pheochromocytomas, is a molecule which plays a critical role in the cross-talk between tumour cells and MCs. Adrenomedullin is an important regulator of MC function related to tumour promotion (Zidaire et al. 2006). Indeed, adrenomedullin induces histamine or β-hexosaminidase release from rat and human MCs through a receptor-independent pathway. It is chemotactic for human MCs and stimulates mRNA expression of VEGF, MCP-1 and FGF-2 in these cells. Immunohistochemical analysis has identified adrenomedullin-producing MCs in tumour infiltrate of human breast and lung cancer patients. In mixed culture assay, the adrenomedullin-producing human MC line HMC-1 augments both anchorage-dependent and -independent growth of human lung cancer A549 cells, an effect suppressed by MC-targeted siRNA adrenomedullin knockdown. Finally, HMC-1 cells induce *in vivo* angiogenesis as assessed by direct *in vivo* angiogenesis assay analysis; neutralizing anti-adrenomedullin monoclonal antibody blocks this ability. Thus, adrenomedullin appears to be implicated as a cross-talk molecule in tumour and MC communication during cancer promotion.

Renin, the rate-limiting enzyme in the activation of the renin-angiotensin system, is synthesized and stored in cardiac mast cells (Mackins et al. 2006). Renin is responsible for angiotensin I formation. Local generation of cardiac angiotensin II from MC-derived renin also elicits noradrenaline release from isolated sympathetic nerve terminals. Another important link has been identified in the heart between sensory nerves and renin-containing MCs. Substance P released from sensory nerves plays a significant role in the release of MC renin in ischemia/reperfusion and in the activation of a local cardiac renin-angiotensin system (Morrey et al. 2010). This culminates in angiotensin production, noradrenaline release, and arrhythmic cardiac dysfunction. Renin is also stored and released by lung MCs (Veerappan et al. 2008). Recently, the existence of an airway renin-angiotensin system triggered by release of MC renin has been demonstrated. In addition, data show that locally produced angiotensin II is a critical factor governing bronchoconstriction.

It has been recently discovered that human MCs secrete amphiregulin after aggregation of FcεRI (Okumura et al. 2005). Amphiregulin is a cytokine of the epidermal growth factor (EGF) family which upregulates mucin gene expression in airway epithelial cells. Upregulation of amphiregulin expression was observed in MCs of patients with asthma, but not normal control subjects. Furthermore, upregulation of amphiregulin in MCs significantly correlates with the extent of goblet cell hyperplasia in the mucosa of patients with bronchial asthma. Amphiregulin secreted by human cord blood-derived MCs induces proliferation of primary human lung fibroblasts (Wang et al. 2005).

A significant increase of mRNA expression of thymic stromal lymphopoietin (TSLP), an IL-7-like cytokine that is capable of triggering dendritic cell-mediated Th2 inflammatory responses, has been reported in the bronchial MCs of asthmatic subjects (Okayama et al. 2009). FcεRI-mediated activation of human MCs upregulates TSLP mRNA expression. In addition, pre-incubation of MCs with IL-4 for 48 h significantly enhances the FcεRI-mediated TSLP mRNA expression and the amount of TSLP in the cell pellets increased significantly. Thus, MCs are able to store TSLP intracellularly and to produce TSLP following aggregation of FcεRI in the presence of IL-4.

Activin A, a homodimeric protein member of the TGF-β superfamily involved in the inflammatory repair process, has been shown to co-localize with human lung MCs from patients with asthma by double-immunofluorescence staining and is secreted by MCs in the airways of a murine asthma model (Cho et al. 2003). This suggests that MC-derived activin A may play an important role in the process of airway remodelling by promoting the proliferation of airway smooth muscles.

MCs secrete different antimicrobial peptides. The cathelicidin LL-37 represents a potent antimicrobial and cell-stimulating agent, most abundantly expressed in peripheral organs such as lung and skin during inflammation. Cultured murine MCs contain abundant amounts of cathelin-related antimicrobial peptide, the murine cath, and this expression is inducible by LPS or lipoteichoic acid (Di Nardo et al. 2003). Human skin MCs also express cath as detected by immunohistochemical analysis for the human cath LL-37. Released LL-37 displays antimicrobial activity against a broad spectrum of microorganisms, neutralizes LPS bioactivity, and chemoattracts neutrophils, monocytes, MCs, and T cells. LL-37 activates airway epithelial cells (Tjabringa et al. 2003). Interestingly, physiological concentrations of LL-37 induce degranulation in purified human lung MCs. As a consequence, LL-37 rapidly undergoes limited cleavage by a released β-tryptase (Schiemann et al. 2009). In addition, LL-37 increases the level of TLR4 mRNA and TLR4 protein, and induces the release of IL-4, IL-5 and IL-1β from cutaneous MCs (Yoshioka et al. 2008). LL-37 co-existing with the bacterial component switches MC function and directs human MCs toward innate immunity. It has recently been demonstrated that the antimicrobial peptides human β-defensins and cathelicidin LL-37 induce the secretion of a pruritogenic cytokine IL-31 by human MCs (Niyonsaba et al. 2010).

## 3.6 Chemokines

MCs have been shown to express an important set of chemokines, which influence recruitment of dendritic cells, lymphocytes, other inflammatory cells and tissue resident cells at sites of tissue inflammation as well as migration of dendritic cells to lymph nodes. Upon TLR3 activation, MCs release chemokines such as MIP-1β, keratinocyte-derived chemokine (KC) and CCL5 (RANTES) (Orinska et al. 2005). Cross-linking of FcεRI on MCs induces the release of the CCL1 chemokine, which acts to recruit Langerhans-type dendritic cells to sites of atopic skin inflammation (Gombert et al. 2005). MCs also express CCL19, the ligand for CCR7, a chemokine receptor required for dendritic cell migration (Humrich et al. 2006). Other products of MC activation, including CCL5 and TNF-α, can promote dendritic cell migration (Yamazaki et al. 1998). MCs also appear to orchestrate the migration of T cells. Upon activation, MCs express chemoattractants such as $LTB_4$, IL-16, CXCL1 (also called lymphotactin), CCL3 (also called MIP-1α), CCL2 (mMCP-1), CCL5, CXCL10, CCL19, and CCL21 (Galli et al. 2005b; Sayed et al. 2008). MC-derived $LTB_4$ is essential for the recruitment of both $CD4^+$ and $CD8^+$ effector cells to sites of inflammation (Ott et al. 2003). CXCL8 (IL-8) is another MC-derived chemokine which exerts basic functions on inflammatory cell recruitment and endothelial cell activation. Platelet factor-4 (PF4) or CXCL4 is a 70-amino acid protein that is released from the alpha-granules of activated platelets and binds with high affinity to heparin. Its major physiologic role appears to be neutralization of heparin-like molecules on the endothelial surface of blood vessels, thereby inhibiting local antithrombin III activity and promoting coagulation. As a strong chemoattractant for neutrophils and fibroblasts, PF4 probably has a role in inflammation and wound repair (Eisman et al. 1990). Immunohistochemical localization by electron microscopy of human PF4 in tissue MCs indicates that this protein is present in MC granules (McLaren and Pepper 1983).

# Chapter 4
# The Tumour Cell

## 4.1 Biology of Tumour Cell

The basic trait of cancer cell is its property to break the social ecosystem and the commitment to reciprocal collaboration of multicellular living organisms. In this perspective, cancer may be defined as "a disease in which an individual mutant clone of cells begins by prospering at the expense of its neighbors" (Alberts et al. 2008). The main implication of this assertion is that tumour cells have the two-fold tendency to escape division control and invade neighbouring tissues as well as colonize at distance. When they merely undergo local expansion, they are classified as benign tumours; but when they infiltrate neighbouring structures, in particular blood vessels and lymphatic channels, they are called malignant tumours. Thus, malignant tumours metastatize and disseminate at distance forming new tumour colonies, which potentially lead to host destruction. The original tumour is defined as the primary tumour. Propagation at distance is called the secondary tumour. Tumours may derive from different kind of tissue cells. Most human cancers are carcinomas, that is tumours which derive from epithelial tissues. Tumours of connective tissue origin, the sarcomas, and tumours arising from white blood cells and haemopoietic cells, the leukemias and lymphomas, are less frequent.

Current wisdom is that cancers derive from a single abnormal cell which has undergone a series of heritable somatic mutations through a progression of molecular events called carcinogenesis. A fundamental aspect of carcinogenesis is the process of mutagenesis, the mechanism which is responsible for modification of the normal DNA sequence. This process typically occurs in chemical carcinogenesis and tumour development by radiation. A single gene mutation, however, is not sufficient for generating tumour cells. Multiple genetic mutations in a single cell are needed to promote aberrant cell behaviour and development of cancer. Thus, stable genetic changes which accumulate in the time allow the abnormal cell to progressively acquire independence from restraining signals derived from neighbouring cells. Abnormalities of the DNA sequence are experimentally detectable in transformed cells. There are numerous and chemically different agents that can cause cancer, known as carcinogens. Many of them are aromatic hydrocarbons and derivatives such as aromatic amines, nitrosamines and alkylating agents. They damage DNA and so

D. Ribatti, E. Crivellato, *Mast Cells and Tumours*,
DOI 10.1007/978-94-007-1469-4_4, © Springer Science+Business Media B.V. 2011

generate mutations. Substances are regarded as tumour initiators or tumour promoters according to their capacity to induce tumour transformation or to facilitate this process. Interestingly, some human cancers are thought to arise by mechanisms that involve viruses, in particular the DNA viruses, such hepatitis-B and -C viruses, human immunodeficiency (HIV) viruses, human herpes virus (HHV) and papilloma viruses. Molecular biology investigation has identified a series of "cancer-critical genes". Alterations of these genes frequently contribute to the causation of tumours. Many cancer-critical genes regulate cell proliferation by acting at different levels of the cell-cycle progression. Indeed, distinct pathways may mediate the disregulation of cell-cycle progression and the disregulation of cell growth in cancer cells. There are two classes of cancer-critical genes. Genes of the first class are called proto-oncogenes. A gain-of-function mutation in these genes can drive a cell toward cancer. Oncogenes are called the mutant, overactive or overexpressed forms of these genes. Genes of the second class are called tumour suppressor genes. A loss-of-function mutation in these genes can contribute to cancer. For instance, DNA tumour viruses block the action of key tumour suppressor proteins. Remarkably, both genetic and epigenetic mechanisms can inactivate tumour suppressor genes. Indeed, it should be added that concomitant epigenetic changes are also involved in the development of cancer. They consist in a profound alteration of the proteins that determine chromatin structures, which leads to a distinct state of genes expression, of the type that occurs in differentiating cells during embryo development. Epigenetic changes involve heterochromatin formation which, at the molecular level, is associated with silencing of specific genes that are adverse to tumour progression.

Recent investigation has provided insight into the concept that many tumours are maintained by a small population of cancer stem cells. The great bulk of the tumour mass, however, is formed by the so called "cancer transit amplifying cells". According to this hierarchical organization, there are a few cancer stem cells in a given tumour which are capable of indefinite self-renewal as well as the production of rapidly dividing cells that accomplish the main tumour mass. The main implication of this crucial discovery is that many tumour stem cells originate from normal tissue stem cells which have accumulated the structural mutations and the epigenetic changes responsible for tumour induction. Conversely, some other tumour stem cells derive from a more differentiated and proliferating cell population inside the original tissue.

One of the most detrimental aspects of tumour behaviour is the capacity of cancer cells to colonize body sites distant from the original place of tumour growth. This process is called "metastasis". It is made possible by a complex series of biological events consisting in the local destruction and tissue invasion, penetration into the blood and lymphatic vessels, dissemination at distant sites and the establishment of tumour micrometastasis. Metastatic events are rare. The great majority of such micrometastasis has little probability to survive to cell apoptosis but some of them show the capability to proliferate extensively giving rise to clinically relevant metastatic colonies. The main way to accomplish tumour metastases is by tumour cell spreading through blood and lymphatic vessels. Tumour cells have the capacity to induce angiogenesis and lymphangiogenesis. Tumours need adequate blood supply for their growth. Thus, tumour cells release angiogenic mediators which recruit blood vessels from preexisting vascular networks. Newly formed blood vessels are structurally deficient and provide easy escape routes that allow for the passage of tumour cells spreading at distance.

The tumour mass is composed not only by tumor cells but it contains a variety of other cell types, which contribute to the establishment of the tumour microenvironment. Non-tumour cell types mixed with tumour cells form the so-called "tumour stroma". It is made-up of fibroblasts of the tumour-associated connective tissue, of blood and lymphatic vessels, and of inflammatory cells. It must be stressed that cells of the tumor stroma are not neoplastic cells although recent work has demonstrated that some altered genetic profiles and phenotypic expression are detectable in endothelial cells of tumour-associated blood vessels. The cross-talk between tumour cells and stromal cells has crucial implications for tumour growth. As illustrated in the following heading of this chapter, the tumour microenvironment provides essential clues for either tumour progression or tumour suppression. In this context, the discovery of the presence of MCs in the tumour stroma is of great significance for our understanding of the mechanisms underlying tissue invasion by tumour cells and their escape through blood and lymphatic vessels.

## 4.2 The Importance of Microenvironment in Tumour Metastasis

It has long been accepted that most malignant tumours show an organ-specific pattern of metastasis. For example, colon carcinomas metastasize usually to liver and lung but rarely to bone, skin or brain and almost never to kidneys, intestine or muscle. In contrast, other tumour entities, such as breast carcinomas, frequently form metastases in most of these organs. This specific formation of secondary tumours at distant sites appears to require a number of steps that must be successfully completed by metastazing tumour cells (Chambers et al. 2002).

Various explanations have been proposed for the site selectivity of blood-bone metastases, including tumour cell surface characteristics (Reading and Hutchins 1985; Turner 1982), response to organ-derived chemotactic factors (Hujanen and Terranova 1985), adhesion between tumour cells and the target organ components (Nicolson 1988a, b) and response to specific host tissue growth factors (Nicolson and Dulski 1986). The relative importance of pre-existing tumour subpopulations with specific metastatic properties and the organ environment characteristics in determining metastatic homing have been debated (Fidler 1986; Talmadge and Fidler 1982; Weiss 1979).

An alternative explanation for the different sites of tumour growth involves interactions between the metastatic cells and the organ environment, possibly in terms of specific binding to endothelial cells and responses to local growth factors. Endothelial cells in the vasculature of different organs express different cell-surface receptors and growth factors that influence the phenotype of the corresponding metastases. Greene and Harvey (1964) first suggested that the organ distribution patterns of metastatic foci were dependent on the formation of sufficient adhesive bonds between arrested tumour cells and endothelial cells, and they hypothesized that these interactions were similar to lymphocyte/endothelial cells at sites of inflammation.

The development of organ-derived microvascular endothelial cell cultures has allowed more specific studies on the preferential homing of tumour cells. Auerbach et al. (1985, 1987) found that different tumours showed differences in their adhesive propensity and preference for different endothelial cells, and in a few cases preferential adhesion was observed to the endothelial cells derived from the organ of origin and the target organ.

## 4.3 The "Seed and Soil" Theory

In 1889, the English surgeon Stephen Paget published his "seed and soil" explanation of the nonrandom pattern of metastasis, and was the first to suggest that interactions between tumour cells and host cells in the microenvironment are critical in regulating tumorigenesis (Ribatti et al. 2006) (Fig. 4.1). Certain favoured tumour

**Fig.4.1** A portrait of Stephen Paget

cells (the "seed"), he said, had a special affinity for the growth-enhancing milieu within specific organs (the "soil"), and hence metastasis only occurred when the "seed" and the "soil" were compatible.

The importance of several components of the "soil" in regulating tumour growth has since been emphasised: (1) the extracellular matrix; (2) stromal cells and their growth factors and inhibitors; (3) microvessels and angiogenic factors; (4) inflammatory cells. There is now substantial evidence that tumour growth and progression depend on the crosstalk between malignant cells and their adjacent stromal compartment.

Experimental data to support Paget's 'seed and soil' hypothesis were derived from studies provided by Fidler and Kriple (1977) using the mouse B16 melanoma cells. They showed that different tumour cell clones, each derived from individual cells isolated from a parent tumour, vary markedly in their ability to form pulmonary nodules following intravenous inoculation of B16 melanoma cells into syngeneic C57BL/6 mice. Tumour growth developed in the lungs and in fragments of pulmonary or ovarian tissue that were implanted intramuscularly. By contrast, metastatic lesions did not develop in implanted renal tissue, or at the site of surgical trauma.

A detailed analysis of experimental metastasis in syngeneic mice indicated that mechanical arrest of tumour cells in the capillary bed of distant organs could indeed occur, but that subsequent proliferation and growth into secondary lesions was influenced by specific organ cells (Hart and Fidler 1980).

Controlled subcloning procedures showed that the observed diversity was not a consequence of the cloning procedures. This indicates that the sites of metastasis are determined not only by the characteristics of the neoplastic cells, but also by the microenvironment of the host tissue (Hart and Fidler 1980).

To exclude the possibility that the metastatic heterogeneity of B16 melanoma cells might have been introduced as a result of the lengthy cultivation, studies on the biological and metastatic heterogeneity of spontaneous tumours were carried out. Melanomas were induced in mice by chronic exposure to ultraviolet B-irradiation and the tumour-promoting agent Croton oil, and tumour metastases were found to differ greatly from each other and from the parent tumour. In addition to differences in the number of metastases that developed from each tumour, there was also significant variability in the size and pigmentation of the metastases. Metastases to the lymph nodes, brain, heart, liver and skin were found in addition to lung metastases. Those growing in the brain were uniformly pigmented, whereas those growing in other organs generally were not (Fidler et al. 1981).

Other observations relating the 'seed and soil' hypothesis were made by Pilgrim (1969), using a transplantable reticulum cell sarcoma, which selectively metastasized to the mouse spleen. When equal numbers of cells were injected into the kidney and the spleen, growth in the spleen was always considerably greater than in the kidney. However, in no case was the mitotic index higher in the spleen than in the kidney. Pilgrim therefore considered that cell loss in the kidney was greater than in the spleen; however, his emphasis was on cell migration rather than cell death within the target organ. Regardless of mechanism, compared with the spleen, the kidney was therefore unfavourable 'soil' for this tumour.

Paget postulated that microenvironment provides a fertile 'soil' for cancer cells endowed with a capacity to grow under specific conditions provided by the 'soil'. A current definition of the 'seed and soil' hypothesis consists of three principles. First, neoplasms are biologically heterogeneous and contain subpopulations of cells with different angiogenic, invasive and metastatic properties. Second, the process of metastasis is selective for cells that succeed in invasion, embolization, survival in the circulation, arrest in a distant capillary bed, and extravasation into and multiplication within the organ parenchyma. Third, the outcome of metastasis depends on multiple interactions of metastatic cells with homeostatic mechanisms, which the tumour cells can escape.

# Chapter 5
# Tumour Angiogenesis

In 1971, Judah Folkman first advanced the hypothesis that tumour growth depends on the formation of new blood vessels from the preexisting vascular bed (Folkman 1971). Angiogenesis and the production of angiogenic factors are fundamental for tumour progression in the form of growth, invasion and metastasis.

Solid tumour growth occurs by means of an avascular phase followed by a vascular phase. Assuming that such growth is dependent on angiogenesis and that this depends on the release of angiogenic factors, the acquisition of an angiogenic ability can be seen as an expression of progression from neoplastic transformation to tumour growth and metastasis (Ribatti et al. 1999a).

The avascular phase appears to correspond to the histopathological picture presented by a small colony of neoplastic cells that reaches a steady state before it proliferates and becomes rapidly invasive. In this scenario, metabolites and catabolites are transferred by simple diffusion through the surrounding tissue. The cells at the periphery of the tumour continue to reproduce, whereas those in the deeper portion die away. Dormant tumours have been discovered during autopsies of individuals who died of causes other than cancer (Black and Welch 1993). Carcinoma *in situ* is found in 98% of individuals aged 50 to 70 years who died of trauma, but is diagnosed in only 0.1% during life. Malignant tumours can grow beyond the critical size of 2 mm at their site of origin by exploiting the host's preexisting vessels. This occurs in tumours implanted in the rat brain (Holash et al. 1999a) and in naturally occurring human lung carcinomas (Pezzella et al. 1997). These finding support the notion that only a very small subset of dormant tumours enters the vascular phase.

Practically all solid tumours, including those of the colon, lung, breast, cervix, bladder, prostate and pancreas, progress through these two phases. The role of angiogenesis in the growth and survival of leukemias and other haematological malignancies has only become evident since 1994 thanks to a series of studies demonstrating that progression in several forms is clearly related to their degree of angiogenesis (Vacca and Ribatti 2006).

D. Ribatti, E. Crivellato, *Mast Cells and Tumours*,

DOI 10.1007/978-94-007-1469-4_5, © Springer Science+Business Media B.V. 2011

## 5.1 The Angiogenic Switch

The angiogenic switch whereby the normally quiescent vasculature give rise to new capillaries separates the avascular (prevascular) phase characterized by a dormant tumour and the vascular phase in which exponential tumour growth ensues (Ribatti et al. 2007b). In the prevascular phase, tumour cells proliferate (sometimes as rapidly as in the vascularized tumour), but the rate of tumour cell death (apoptosis) counterbalances this proliferation and maintains the tumour mass in a steady state. Dormant tumours have been discovered during autopsies of individuals who died for other causes (Kirsch et al. 2004).

Activation of the switch itself has been attributed to the synthesis or release of angiogenic factors. The balance hypothesis (Hanahan and Folkman 1996; Bouck et al. 1996) assumes that the level of angiogenesis inducers and inhibitors governs cell differentiation states of quiescence or angiogenesis. This balance is altered by increasing activator gene expression, changing the bioavailability or activity of the inducer proteins, or reducing the concentrations of endogenous angiogenesis inhibitors, here, too, via changes in gene expression or processing/biovailability.

The switch depends on increased production of one or more of the positive regulators of angiogenesis, such as VEGF, FGF-2, IL-8, TGF-β, PDGF, pleiotrophins and others. These can be exported from tumour cells (Kandel et al. 1991), mobilized from the extracellular matrix (Vlodavsky et al. 1990) or released from host cells (e.g. macrophages) recruited to the tumour (Leibovich et al. 1987). Expression of endogenous inhibitors, such as thrombospondin-1 (TSP-1) or insulin-like growth factor (IGF-β) may be downregulated (Rastinejad et al. 1989; Good et al. 1990; Bornstein 1992; Dameron et al. 1994). Thus, the switch clearly involves more than simple upregulation of angiogenic activity and has thus been seen as the result of a net balance of positive and negative regulators.

Integrin signalling also contributes to this regulatory balance. Quiescent vessels express one class of integrins, whereas sprouting capillaries express another. Interference with signaling by the latter class of integrins can inhibit angiogenesis (Giancotti and Ruoslahti 1999). *In vivo* screening of phage libraries in murine models has identified specific motifs, including RGD, GSL, and NGR, that bind to integrins αvβ3, αvβ5 and α5β1, MMPs and VEGFR that are upregulated in neoangiogenic tumour or endothelial cells (Pasqualini and Arap 2002; Ruoslahti 2002).

Proteases control the bioavailability of angiogenic activators and inhibitors. Some release FGF-2 stored in the extracellular matrix (Whitelock et al. 1996), whereas plasmin, a proangiogenic component of the clotting system, cleaves itself into an angiogenesis inhibitor form, namely angiostatin (Gately et al. 1997).

Nutrient deprivation modulates gene expression and may also contribute to the activation of the angiogenic process. Glucose deprivation-induced oxidative stress activates the expression or release of angiogenic growth factors (Spitz 2000).

The mechanism of the switch was first described by Hanahan (1985) who developed transgenic mice in which the large T oncogene is hybridized to the insulin promoter. In this islet cell tumorigenesis (RIP1-TAG2 model), these mice express the large T antigen in all their islet cells at birth, and express the SV40 T antigen (TAG) under the control of the insulin gene promoter, which elicits the sequential development of tumours in the islets over a period of 12–14 weeks. Tumour development proceeds by stages during which about half the 400 islets hyperproliferate, while a subset (about 25%) subsequently acquire the ability to switch to angiogenesis (Folkman et al. 1989). Some 15–20% of these angiogenic islets develop into benign tumours, encapsulated lesions and invasive carcinomas (Lopez and Hanahan 2002). This multistage pathway suggests the sequential involvement of multiple rate-limiting genetic and epigenetic events in the progression from normal cells to tumours. The β-cells become hyperplastic and progress to tumours via a reproducible and predictable multistep process. One step occurs at 6–7 weeks, when angiogenesis is switched on in approximately 10% of preneoplastic islets. Solid vascularized tumours first appear at 9–10 weeks, initially as small nodules that grow and progress to large islet tumours, with well-defined margins, as well as two classes of invasive carcinoma (Lopez and Hanahan 2002). Lopez and Hanahan (2002) identified stage-specific molecular markers accessible via the circulation, either on the surface of endothelial cells, their peri-endothelial support cells (pericytes and smooth muscle cells) or even tumour cells themselves (as a result of the hemorrhagic leaky angiogenic vasculature). They selected phage pools that homed preferentially to different stages during RIP1-TAG2 tumorigenesis. In addition to 'panangiogenic' markers shared by many types of tumours, they identified vascular target molecules characteristic of this tumour's tissue of origin and not expressed in the vessels of several tumour types growing in or under the skin.

Angiogenic islets are revealed both morphologically in tissue sections and in isolated islets by their red colour and microhemorrhagic islands, and functionally by their ability to elicit endothelial cell migration, proliferation and tube formation in an *in vitro* collagen bioassay involving co-culture of dispersed capillary endothelial cells and isolated islets (Folkman et al. 1989). This onset pattern closely resembles that of angiogenesis in human tumours.

Two concepts emerged from this early characterization of tumorigenesis in RIP-TAG transgenic mice: (1) the existence of distinct stages of premalignant progression, namely a hyperplastic stage followed by a stochastic angiogenic stage and (2) the development of angiogenesis well before the emergence of an invasive malignancy. The temporal and histological changes that occur in the RIP-TAG model are consistent with the multistep paradigm for tumorigenesis of human cancers (Vogelstein 1993). The high incidence of occult human cancers suggests that this angiogenic switch, as in the RIP-TAG model, may be a relatively late event that plays a significant role in the transition from microscopic foci to macroscopic tumour (Udagawa et al. 2002).

These data suggested that induction of angiogenesis during multistage carcinogenesis is coordinated by an "angiogenic switch". VEGF signalling is primarily implicated in angiogenesis and tumorigenesis in RIP1-TAG2 mice. The islets are extensively vascularized to facilitate their monitoring of serum glucose levels and hence the secretion of insulin and other hormones for endocrine regulation of carbohydrate metabolism. VEGF-A, VEGF-B and VEGF-C are all expressed in normal islet β-cells (Christofori et al. 1995). VEGF-A is expressed at all stages of RIP1-TAG2 tumorigenesis (Christofori et al. 1995). Such constancy suggests that if VEGF-A activity is important in this tumorigenesis pathway, even if other models of regulation may be involved. Inoue et al. (2002) showed that five VEGF ligand genes are expressed in normal islets and throughout tumorigenesis. Moreover, they produced a β-cell-specific VEGF-A knockout that resulted in islets with reduced vascularity, but essentially normal physiology. In RIP1-TAG2 mice where most oncogene-expressing cells had deleted the VEGF-A gene, both angiogenic switching and tumour growth were severely disrupted, as was the neovasculature.

Overexpression of VEGF-C in β-cells via a RIP-VEGF-C transgene induces peri-islet lymphatic vessels, but has no discernable effect on intra-islet blood vessels (Mandriota et al. 2001). Extensive lymphatic channel formation, in fact, was observed around (but not within) 98% of islets from RIP-VEGF-C transgenic mice, i.e. the anatomical units in which the transgene is expressed. VEGF-C overexpression did not increase intratumoral vascularity or enhance tumour growth, though peri-tumoral lymphatic vessels it induced facilitated metastases to the draining mesenteric lymph nodes (Mandriota et al. 2001). To assess the role of VEGF-C-induced lymphangiogenesis in tumour metastasis, Mandriota et al. (2001) crossed RIP-VEGF-C mice with RIP1-TAG2 mice. β-cell tumours in RIP1-TAG2 mice invade locally, but are not metastatic. VEGF-C expression in double transgenics resulted in the *de novo* formation of lymphatics in intimate association with β-cell tumors associated with the formation of metastases in the draining regional mesenteric lymph nodes in 37% of mice.

Mice genetically deficient in PDGFB or its receptors have blood vessels with loose pericyte attachment, irregular vessel caliber, luminal projections of endothelial cells and haemorrhage (Betsholtz 2004). Similar abnormalities occur in many tumour vessels.

Bergers et al. (2000) demonstrated that MMP-9 plays a crucial role in the initial angiogenic switch during islet carcinogenesis and proposed mobilization of VEGF from an extracellular reservoir as its mode of action. Giraudo et al. (2004) used the K14-HPV 16 transgenic tumour model to demonstrate MMP-9 in the tumour stroma concomitant with the angiogenic switch, expressed by inflitrating macrophages. Moreover, preclinical trials targeting MMP-9 and angiogenesis with a MMP inhibitor and with a bisphosponate, zoledronic acid, showed that both were antiangiogenic. Other enzymes are involved in islet tumorigenesis. Joyce et al. (2004) have shown that a subset of papain family Clan CA proteases known as cathepsins make an important contribution to the development of islet tumors and are upregulated during their progression. Cathepsin activity was assessed with chemical probes to allow biochemical and *in vivo* imaging. Increased activity was associated with the angiogenic vasculature and invasive fronts of carcinomas, with differential expression in immune, endothelial and cancer cells. A broad-spectrum cysteine inhibitor that knocked out cathepsin function at different stages of tumorigenesis impaired angiogenic switching in progenitor lesions, as well as tumour growth, vascularity and invasiveness. Cysteine cathepsins are also upregulated during HPV16-induced cervical carcinomas. Joyce et al. (2003) have since shown that heparanase expression increases during RIP-TAG tumorigenesis, predominantly supplied by innate immune cell infiltrating neoplastic tissues. Joyce et al. (2003) analyzed the vasculature in the angiogenic stages of RIP-TAG model islet tumorigenesis with phage libraries that display short peptides, and identified peptides that discriminate between the vasculature of the premalignant angiogenic islets and the fully developed vasculature. One peptide is homologous with PDGFB, which is expressed in endothelial cells, while its receptor is expressed in pericytes. Three PDGF ligand genes are expressed in the tumor endothelial cells, while PDGFB-R is expressed in tumour pericytes (Bergers et al. 2003).

## 5.2 Phenotypic and Genotypic Characteristics of Tumour Vessels and Genetic Evidence that Tumours Are Angiogenesis-Dependent

Tumour endothelial cells may divide up to 50 times more frequently than endothelial cells of normal tissues. Considerable differences exist between normal and tumour vasculature. The immaturity of tumour vessels led H. Dvorak to define a tumour as “a wound that never heals” (Dvorak 1986). Although the tumour vasculature originates from the host vessels and the mechanisms of angiogenesis are similar, the organization may differ dramatically depending on the tumour type and its location.

The blood vessels of tumours display many structural and functional abnormalities (Ribatti et al. 2007c). Their unusual leakiness, potential for rapid growth and remodelling, and expression of distinctive surface molecules, mediate the dissemination of tumour cells in the bloodstream and maintain the tumour microenvironment. Like normal blood vessels, they consist of endothelial cells, pericytes and their enveloping basement membrane. Common features, regardless of their origin, size and growth pattern, include the absence of a hierarchy, the formation of large-caliber sinusoidal vessels and a markedly heterogenous density. Low permeability tumours overexpress Ang-1 and/or underexpress VEGF. Conversely, those with high permeability may lack Ang-1 or overexpress its antagononist, Ang-2 (Jain and Munn 2000). Tumour endothelial cells are most markedly activated. Many of the surface markers they express are more strongly expressed by tumoral than normal vessels. For instance, they express much more VEGFR-2, Tie-1, Tie-2, the integrins $\alpha v\beta 3$ and $\alpha v\beta 5$ (Stromblad and Cheresh 1996) or the alternative, spliced variant of fibronectin ED-B (Neri et al. 1997), than resting endothelial cells do in normal tissues. Furthermore, tumour endothelial cells express different levels of adhesion molecules for circulating leukocytes and high levels of E-selectin (Bischoff 1995).

St Croix et al. (2000) compared the gene expression patterns of endothelial cells derived from normal and malignant colorectal tissues. Differential expression was present in 79 transcripts: 46 were elevated at least 10-fold while 33 were expressed at substantially lower levels in tumour-associated endothelial cells. Most of the differentially expressed genes have also been found during luteal angiogenesis and wound healing, which suggests that in tumour angiogenesis the same signalling pathways are involved as in physiologic angiogenesis. These data clearly delineate the effects of the local microenvironment on gene expression patterns in endothelial cells and support the critical role of the microenvironment in defining the angiogenic phenotype.

After the work by St Croix, only a limited number of studies have characterized the gene expression profile of freshly isolated tumour endothelial cells (Madden et al. 2004; Parker et al. 2004). Van Beijnum et al. (2006) compared the transcriptional profiles of angiogenic endothelial cells isolated from both malignant and non-malignant tissues with those of resting endothelial cells and identified 17 genes that show specific overexpression in tumour endothelium, but not in the angiogenic endothelium of normal tissues. Moreover, antibody targeting of four cell-surface expressed or secreted products (vimentin, CD59, HMGB1 and IGFBP7) inhibited angiogenesis *in vitro* and *in vivo.*

It is increasingly recognized that oncogenes, such as mutant RAS or SRC, may also contribute to tumour angiogenesis by enhancing the production of VEGF (Rak et al. 1995; Ellis et al. 1998). Downregulation of the RAS-oncogene in a melanoma driven by doxycycline-inducible RAS led to tumour regression within 12 days (Tang et al. 2005). Cells that expressed low levels of RAS were dormant and non-angiogenic, whereas cells that expressed high levels of RAS developed into full-blown tumours (Watnick et al. 2003). These authors demonstrated that whereas VEGF levels increased only modestly in tumours that expressed high levels of RAS, TSP-1 levels increased markedly in these cells.

Tumours that express bcl-2 escape mitomycin C therapy and grow ~1,000 $mm^3$. When bcl-2-expressing tumors are treated with an angiogenesis inhibitor (TNP-470) which selectively inhibits the proliferation of tumour cells or fibroblasts, the bcl-2 effect is annulled, and tumour growth is restricted to <10–15% of the growth observed in untreated bcl-2 expressing tumours (Fernandez et al. 2001).

## 5.3 What Is the Evidence that Genetic Instability Promotes the Angiogenic Switch?

A widely accepted view is that the progression of tumours reflects their genetic instability, this being defined as their higher mutation rate compared with normal tissues due to corruptions in checkpoint genes crucial for genome replication (Lengauer et al. 1998). Progression is thus achieved through the accumulation of multiple lesions that impair the control of cell proliferation and survival and thereby shape the complex phenotype of tumour cells (Hanahan and Weinberg 2000). Genetic instability may be required for the emergence of angiogenic tumour cell lines that enhance a tumour's growth and malignancy. In the absence of such instability, these cells cannot grow, even if the relevant mutations are generated at low levels because the angiogenic promoters will not sufficient to counter the influence of inhibitory factors. Inhibition of angiogenesis can thus be viewed as a host defense and a tumour must be genetically unstable to be able to exceed a certain size. Genetic control of the physiological levels of endogenous angiogenesis inhibitors may well be a line of defence against the conversion of dormant tumour cells to a malignant phenotype.

Genetic instability must therefore act upstream and promote the angiogenic switch. Evidence in favor of this view has been acquired from reversible transgene models and multigene-transformed cells. Watnick et al. (2003) observed that the switch in a cell transformation model was dependent on oncogenic RAS expression. They showed that low expression levels induced cell transformation and increased VEGF expression, and that further increases in the abundance of the oncogene led to repression of the antiangiogenic factor TSP-1 through Myc activation, and subsequent tumour expansion. Expression of the SV40 early region, TERT and activated RAS were sufficient to transform primary epithelial cells *in vitro* (Dameron et al. 1994). However, their ability to grow *in vivo* depended on the level of RAS expression: cells that expressed low levels were dormant and non-angiogenic, whereas those that expressed high levels progressed to full-blown tumours. VEGF-A levels increased only modestly (1.4 fold) in tumours expressing high RAS levels, whereas TSP1 levels increased 8 fold. Tumour formation from cells that expressed low RAS levels was indicated by simple overexpression of VEGF-A. This shows tumour progression was blocked because the angiogenic switch was not activated. It also supports the view that the switch is determined by the balance between pro- and antiangiogenic factors, and that oncogene expression can influence this balance.

BHK 21/cl 13 cells, an immortal but non-tumorigenic line of hamster fibroblasts, were converted to malignancy and anchorage independence by loss of a functioning tumour suppressor gene. These cells were highly tumorigenic in nude mice and neonatal hamsters, and potently angiogenic *in vivo*. Normal BHK cells and suppressed hybrids generated by fusing transformed BHK cells with either non-transformed BHK or normal human fibroblasts were unable to induce angiogenesis when they or their concentrated conditioned media were introduced into rat corneas, whereas transformed BHK cells and transformed segregants from the suppressed hybrids were angiogenic. Mixing experiments showed that normal cells elaborated an angiogenesis inhibitor whose production was blocked coincidentally with suppressor loss. When endothelial cell chemotaxis was used as an *in vitro* corollary of angiogenesis in the rat cornea assay, the inhibitor was purified and shown to be TSP-1 (Bornstein 1992). This was the first illustration of a new function for a tumour suppressor gene, namely regulation of the production of a naturally occurring inhibitor of angiogenesis. In another set of experiments, Dameron et al. (1994) established a direct link between the p53 tumour suppressor gene, tumour angiogenesis and TSP-1. To examine the effect of p53, they used cultured fibroblasts from patients with the Li-Fraumeni syndrome who have inherited one wild-type allele and one mutant allele of the p53 gene. When the wild-type allele was lost, these cells acquired potent angiogenic activity coincidental with loss of TSP-1 production. Transfection revealed that p53 stimulated the endogenous TSP-1 gene and positively regulated the TSP-1 promoter sequences.

## 5.4 Hypoxic Regulation of Tumour Angiogenesis

There is a complex interrelationship between tumour hypoxia and tumour angiogenesis. Hypoxia in tumours develops in the form of chronic hypoxia, resulting from long diffusion distances between tumour vessels, and/or of acute hypoxia, resulting from a transient collapse of tumour vessels. Many tumours contain a hypoxic microenvironment, a condition that is associated with poor prognosis and resistance to treatment. The production of several angiogenic cytokines, such as FGF-2, VEGF, TGF-β, TNF-α and IL-8, is regulated by hypoxia. VEGF-mRNA expression is rapidly and reversibly induced by exposure of cultured endothelial cells to low $PO_2$ (Levy et al. 1995). Many tumour cell lines have been reported to show hypoxia-induced expression of VEGF (Papetti and Herman 2002; Plate et al. 1992; Shweiki et al. 1992; Potgens et al. 1995; Claffey et al. 1996). In a rat glioma model, VEGF gene expression was activated in a distinct tumour cell subpopulation by two distinct hypoxia-driven mechanisms (Damert et al. 1997). Hypoxia-inducible factor (HIF)-1 helps to restore oxygen homeostasis by inducing glycolysis, erythropoiesis and angiogenesis (Semenza 1996), and tumour vascularization is largely controlled by HIF-1, partly as a result of VEGF upregulation (Carmeliet et al. 1998).

## 5.5 The Role of Pericytes in Tumour Blood Vessels

Among the pathways involved in pericyte recruitment during embryonic development, the contribution of PDGFB is confirmed in tumour angiogenesis. PDGFB expressed by tumour cells increased pericyte recruitment in several *in vivo* tumour models but failed to correct their detachment in PDGFB retention motif deficient mice (Abramsson et al. 2003; Guo et al. 2003). Genetic abolition of the PDGFB receptor expressed by embryonic pericytes decreased their recruitment in tumour (Abramsson et al. 2003). In Lewis lung carcinoma tumours implanted in mice, inhibition by RNA interference of endothelial differentiation gene-1 (EDG-1) expression in endothelial cells strongly reduced pericyte coverage (Chae et al. 2004).

In a human glioma model developed in rat, Ang-1 led to enhanced pericyte recruitment and increased tumour growth, presumably by favouring angiogenesis (Machein et al. 2004). On the contrary, in a colon cancer model, overexpression of Ang-1 led to smaller tumours with fewer blood vessels and a higher degree of pericyte coverage, resulting in a decreased vascular permeability and reduced hepatic metastasis (Ahmad et al. 2001; Stoeltzing et al. 2003). In a human neuroblastoma xenotransplanted model, pericyte coverage along tumour microvessels is decreased by half in tumours grafted to MMP-9 deficient mice, and transplantation with MMP-9 expressing bone marrow cells restores the formation of mature tumour vessels (Chantrain et al. 2004). Overexpression of the tissue inhibitor of MMP-3 (TIMP-3) results in decreased pericyte recruitment in neuroblastoma and melanoma models (Spurbeck et al. 2002).

## 5.6 The Role of Inflammatory Cells in Tumour Angiogenesis

Tumour cells are surrounded by an infiltrate of inflammatory cells, such as lymphocytes, neutrophils, macrophages and MCs. These cells communicate by means of a complex network of intercellular signalling pathways mediated by surface adhesion molecules, cytokines and their receptors (Park et al. 2000). It is becoming clear that stromal cells cooperate with endothelial and cancer cells in promoting angiogenesis, secreting a varied repertoire of growth factors and proteases that enable them to enhance tumour growth.

In breast cancer, release by tumour-associated and tumour-infiltrating neutrophils of oncostatin M, a pleiotropic cytokine belonging to the IL-6 family, promotes tumour progression by enhancing angiogenesis and metastases (Queen et al. 2005). In addition, neutrophil-derived oncostatin M induces VEGF production from cancer cells and increases breast cancer cell detachment and invasive capacity (Coussens and Werb 1996). Expression of HPV 16 early region genes in basal keratinocytes of transgenic mice elicits a multi-stage pathway to squamous carcinoma. Infiltration by neutrophils and MCs, and activation of MMP-9 in these cells coincided with the angiogenic switch in premalignant lesions (Nozawa et al. 2006). In the Rip-Tag2 model of pancreatic islet carcinogenesis, MMP-9-expressing neutrophils were predominantly found in the angiogenic islets of dysplasias and tumours, and transient depletion of neutrophils clearly reduced the frequency of the initial angiogenic switch in the dysplasias (Masson et al. 2005). The lack of both MMP-9-positive neutrophils and MMP-2-expressing stromal cells in mice with a double deficiency for MMP-2 and MMP-9 resulted in a lack of tumour vascularization followed by a lack of tumour invasion (Masson et al. 2005).

Expression of granulocyte-colony stimulating factor (G-CSF) or co-expression of G-CSF and GM-CSF together induced malignant progression of previously benign factor-negative HaCaT tumour cells. This progression was associated with enhanced and accelerated neutrophil recruitment into the tumour vicinity. The neutrophil recruitment preceded the induction of angiogenesis in the HaCaT heterotransplantation model for human squamous cell carcinoma and in nude mouse heterotransplants of head and neck carcinomas (Obermüller et al. 2004; Gutschalk et al. 2006).

In some tumours, like melanoma, neutrophils are not a major constituent of the leukocyte infiltrate, but they might have a key role in triggering and sustaining the inflammatory cascade, providing chemotactic molecules for the recruitment of macrophages and other inflammatory and stromal cells. Neutrophils produce and release high levels of MMP-9. By contrast, neutrophils secrete little, if any, MMP-2, which plays an important role in the turnover of various extracellular matrix components (Muhs et al. 2003). However, neutrophils release a not yet identified soluble factor as well as a specific sulphatase and a heparanase that activate latent MMP-2 secreted by other cells and allow releasing of embedded growth factors from the extracellular matrix (Schwartz et al. 1998). Remodelled matrix facilitates the escape of tumour cells leaving the tumour mass to metastasize at distance, because it offers less resistance. In addition, proteolytic enzymes released by neutrophils can diminish cell-cell interactions and permit the dissociation of tumour cells from the original tumour site (Shamamian et al. 2001).

Cells belonging to the monocyte-macrophage lineage are a major component of the leukocyte infiltration in tumours (Balkwill and Mantovani 2001; Mantovani et al. 2002). A number of tumour-derived chemoattractants ensures macrophage recruitment, including colony-stimulating factor-1 (CSF-1), the CC chemokines CCL2, CCL3, CCL4, CCL5 and CCL8, and VEGF secreted by both tumour and stromal elements (Mantovani et al. 2002). Besides killing tumour cells once activated by IFN-γ and IL-12, tumour-associated macrophages produce several proangiogenic cytokines as well as extracellular matrix-degrading enzymes (Naldini and Carraro 2005). The stimulating effect exerted by tumour-associated macrophages on the growth of the tumour mass is partly related to the angiogenic potential of these cells. In the tumour microenvironment, macrophages are mainly represented by polarized type II (alternatively activated) or M2 elements, which would derive from tumour-associated macrophages upon local exposure to IL-4 and IL-10 (Mantovani et al. 2002). These cells have poor attitude to destroy tumour cells but are better adapted to promoting angiogenesis, repairing and remodelling wounded or damaged tissues, and suppressing adaptive immunity (Sica et al. 2006). Tumour-associated macrophages represent a rich source of potent proangiogenic cytokines and growth factors, such as VEGF, TNF-α, IL-8 and FGF-2. In addition, these cells express a broad array of angiogenesis-modulating enzymes, including MMP-2, -7, -9, -12, and cycloxygenase-2 (COX-2) (Sunderkokker et al. 1991; Lewis et al. 1995; Klimp et al. 2001). In humans, a significant relationship between the number of tumour-associated macrophages and the density of blood vessels has been established in tumours like breast carcinoma (Leek et al. 1996), melanoma (Makitie et al. 2001), glioma (Nishie et al. 1999), squamous cell carcinoma of the esophagus (Koide et al. 2004), bladder carcinoma (Hanada et al. 2000), and prostate carcinoma (Lissbrant et al. 2000). In the mouse cornea model, killing of COX-2 positive infiltrating macrophages with clodronate liposomes reduces IL-1β-induced angiogenesis and partially inhibits VEGF-induced angiogenesis (Nakao et al. 2005). In one model of subcutaneous melanoma, both angiogenesis and growth rate correlate with tumour infiltration by macrophages that express angiotensin I receptor and VEGF (Egami et al. 2003). In addition, Lewis lung carcinoma cells expressing IL-1β develop neovasculature with macrophage infiltration and enhance tumour growth in wild-type but not in MCP-1-deficient mice, suggesting that macrophage involvement might be a prerequisite for IL-1β-induced neovascularization and tumour progression (Nakao et al. 2005) In a murine model of mammary carcinoma, deficiency of macrophage-colony stimulating factor (M-CSF), a potent inductor of macrophage recruitment in tumour tissues, does not affect early stages of tumour development but reduces progression to invasive carcinoma and metastasis (Lin et al. 2001). This result highlights the possible role of tumour-associated macrophages in contributing to the angiogenic switch that accompanies transition into malignancy. In polyoma middle-T (PyMT)-induced mouse mammary tumours, indeed, focal accumulation of macrophages in premalignant lesions precedes the angiogenic switch and the progression into invasive tumours. Depletion of tumour-associated macrophages reduces to about 50% tumour vascular density, leading to areas of necrosis by loss of blood supply within the tumour mass. Interestingly, macrophages have been shown to accumulate particularly in such necrotic and hypoxic areas in different neoplasia, like human endometrial, breast, prostate and ovarian carcinomas (Ohno et al. 2004; Leek et al. 1999). It is otherwise known, indeed, that upregulation of the proangiogenic program in tumour-associated macrophages, followed by increased release of VEGF, FGF-2, TNF-α, urokinase and MMPs, is stimulated by hypoxia and acidosis (Bingle et al. 2002). Moreover, activated macrophages synthesize and release inducible nitric oxide synthase, which increases blood flow and promotes angiogenesis (Jenkins et al. 1995). Lastly, the angiogenic factors secreted by macrophages stimulate migration of other accessory cells that potentiate angiogenesis, in particular MCs (Gruber et al. 1995). Osteopontin deeply affects the proangiogenic potential of human monocytes (Denhardt et al. 2001). Reports suggest that osteopontin may affect angiogenesis by acting directly on endothelial cells and/or indirectly via mononuclear phagocyte engagement, enhancing the expression of TNF-α and IL-1β in mononuclear cells (Leali et al. 2003; Naldini et al. 2006).

It should also be mentioned that monocytes and macrophages are primary producers of IL-12. This multifunctional cytokine can cause tumour regression and reduce metastasis in animal models, due to the promotion of anti-tumour immunity and also to the significant inhibition of angiogenesis (Colombo and Trincheri 2002). The antiangiogenic activity is mediated by IFN-γ production, which in turn induces the chemokine IFN-γ-inducible protein-10 (Angiolillo et al. 1995; Romagnani et al. 2001). There is *in vitro* evidence that IL-12 inhibits VEGF produced by breast cancer cells and regulates stromal cell interactions, leading to decreased MMP-9 and increased TIMP-1 production (Dias et al. 1998).

Tumour-associated macrophages accumulate in poorly vascularized hypoxic or necrotic areas (Leek et al. 1999), and respond to experimental hypoxia by increasing the release of VEGF and FGF-2 and a broad range of other factors, such as PDGF, TNF-α, FGF-2, VEGF, uPA and MMPs (Bingle et al. 2002). Moreover, activated macrophages synthesize and release inducible nitric oxide synthase (NOS), which increases blood flow and promotes angiogenesis (Jenkins et al. 1995). The angiogenic factors secreted by macrophages stimulate MC migration (Gruber et al. 1995). Lin and Pollard (2004) showed that in the PyMT model of mammary carcinogenesis, macrophages are recruited to premalignant tumours immediately before the angiogenic switch that precedes the transition to a malignant phenotype. Depletion of these macrophages resulted in a decreased vascular density in tumours.

Monocytes-macrophages and endothelial cells share phenotypical and functional features, including the expression of common metabolic and surface markers, as well as the ability to transdifferentiate into endothelial cells *in vitro* and *in vivo* (Fernandez Pujol et al. 2000; Schmeisser et al. 2001; Iba et al. 2002; Anghelina et al. 2004). Venneri et al. (2007) have reported the identification in human peripheral blood of a novel subset of Tie-2 expressing monocytes (TEMs) that promote angiogenesis in paracrine manner. Although recruited to tumours in lower numbers than tumour associated macrophages, TEMs are a more potent source of proangiogenic signals, suggesting that they significantly contribute to tumour angiogenesis.

Gottfried et al. (2007) demonstrated that incubation of tumour-associated dendritic cells with VEGF and oncostatin M led to transdifferentation into endothelial-like cells. These cells showed strong expression of classical endothelial cell markers, such as von Willebrand factor and vascular endothelial cadherin, while leukocytic markers were reduced. Moreover, they were able to induce vascular-like tubes on Matrigel.

## 5.7 Participation of Haematopoietic Cells and Endothelial Precursor Cells to Tumour Angiogenesis

High levels of VEGF produced by tumours may result in mobilization of endothelial precursor cells (EPCs) in the peripheral circulation and enhance their recruitment into the tumour vasculature (Asahara et al. 1999). Hypoxia mobilizes EPCs from the bone marrow in the same way as it does haematopoietic cytokines, such as GM-CSF (Takahashi et al. 1999). Malignant tumour growth results in neoplastic tissue hypoxia, and may mobilize bone-marrow-derived EPCs in a paracrine fashion and thus contribute to promote the sprouting of new vessels.

By introducing a suicide gene with a lentiviral vector, De Palma et al. (2003a, b) conditionally eliminated bone marrow-derived haematopoietic cells (HCs) during the early phases of tumour growth and showed that the impaired recruitment of these cells inhibited tumour angiogenesis and progression.

Carcinoma-associated fibroblasts isolated from breast cancer secrete high levels of stromal cell derived factor-1 (SDF-1), a molecule that functions as a potent chemoattractant for endothelial cells and HCs and enhances tumour angiogenesis (Orimo et al. 2005).

The importance of EPCs mobilization in tumour vascularization has been demonstrated in a mouse mutant for Id proteins (Lyden et al. 2001). Tumours xenografted in lethally irradiated Id1+/−Id3−/− mice and reconstituted with wild type EPCs grow as well in control mice and most of the endothelial cells are also Id3+. Moreover, in wild type mice, EPC recruitment to the tumour vasculature is completely inhibited by the injection of neutralizing antibody against VEGFR-2, but not against VEGFR-1.

The extent of EPC recruitment into the tumour vasculature may depend on the tumour type, varying from 90% in lymphoma to 5% in neuroblastoma implanted subcutaneously in mice (Davidoff et al. 2001).

## 5.8 Alternative Mechanisms of Tumour Angiogenesis

Maniotis et al. (1999) described a new model of formation of vascular channels by human melanoma cells, and called it "vasculogenic mimicry" to emphasize the *de novo* generation of blood vessels without the participation of endothelial cells and independent of angiogenesis. The word "vasculogenic" was selected to indicate the *de novo* generation of the pathway and "mimicry" was used because tumour cell pathways serving to transport fluid to tissues were clearly not blood vessels.

Microarray gene chip analysis of highly aggressive human cutaneous melanoma cell lines compared with poorly aggressive lines revealed a significant increase in the expression of laminin 5 and MMP-1, -2, and -9 and MT1-MMP in the highly aggressive cells (Seftor et al. 2001), suggesting that they interact with and alter their extracellular environment in a different way as compared with poorly aggressive cells, and that increased expression of MMP-2 and MT1-MMP, along with matrix deposition of laminin 5, are required to achieve their "vasculogenic mimicry". These data have been vigorously disputed by McDonald et al. (2000), who consider that the evidence presented is neither persuasive nor novel. In their opinion, the data are not convincing because three key questions were not addressed: (1) if erythrocytes are used as markers, are they located inside or outside blood vessels?; (2) where is the interface between endothelial cells and tumour cells in the blood vessel wall?; (3) how extensive is the presumptive contribution of tumour cells to the lining of blood vessels?

Another possibility is that the endothelial cells lining is replaced by tumour cells, resulting in so-called "mosaic vessels", where both endothelial and tumour cells contribute to form of the vascular tube (Chang et al. 2000). These Authors used CD31 and CD105 to identify endothelial cells and endogenous green fluorescent protein (GFP) labeling of tumour cells, and showed that approximately 15% of perfused vessels of a colon carcinoma xenografted at two sites in mice were mosaic, with focal regions where no CD31/CD105 immnoreactivity was detected and tumour cells were in contact with the vessel lumen.

## 5.9 Vascular Cooption

Holash et al. (1999a) reported that tumour cells migrate toward existing host organ blood vessels in sites of metastases, or in vascularized organs such as the brain, to trigger blood vessel-dependent tumour growth as opposed to classic angiogenesis. These vessels then regress owing to apoptosis of the constituent endothelial cells, apparently mediated by Ang-2. Finally, at the periphery of the growing tumour mass angiogenesis occurs by the cooperative interaction of VEGF and Ang-2. Tumour cells often appear to have immediate access to blood vessels, such as when they metastasize to or are implanted within a vascularized tissue (Holash et al. 1999a; Zagzag et al. 1999). They immediately coopt existing adjacent vessels and often grow as cuffs around them. A robust host defence mechanism is activated, in which the coopted vessels initiate an apoptotic cascade, probably by autocrine induction of Ang-2, followed by regression of the coopted vessels, that carries off much of the dependent tumour and results in massive tumour death. However, successful tumours overcome this vessel regression by initiating neo-angiogenesis.

Many solid tumours may fail to form a well-differentiated and stable vasculature because their newly formed tumour vessels continue to overexpress Ang-2. Ang-2 induction in host vessels in the periphery of experimental C6 glioma precedes VEGF upregulation of tumour cells, and causes regression of coopted vessels (Holash et al. 1999b; Yancopoulos et al. 2000).

Vajkoczy et al. (2002) have demonstrated a parallel induction of Ang-2 and VEGFR-2 in quiescent host endothelial cells, suggesting that their simultaneous activity is critical for the induction of tumour angiogenesis during vascular initiation of microtumours. Consequently, the simultaneous expression of VEGFR-2 and Ang-2, rather than the expression of Ang-2 alone, may indicate the endothelial cell angiogenic phenotype and thus provide an early marker of activated host vasculature. The VEGF/Ang-2 balance may determine whether the new tumour vessels will continue to expand when the ratio of VEGF to Ang-2 is high, or regress when it is low during remodelling of the tumour microvasculature.

## 5.10 The Prognostic Significance of Tumour Angiogenesis

In 1991, Weidner and co-workers used specific antiendothelial antibodies to highlight tumour vasculature, in order to demonstrate that microvessel density was a prognostic marker for human breast cancer (Weidner et al. 1991). Since then, the majority of reports have confirmed that microvessel density is a powerful and often independent prognostic indicator for many different types of human cancers, such as breast cancer, prostate cancer, melanoma, ovarian carcinoma, gastric carcinoma and colon carcinoma. However, a few other reports failed to show that microvessel density is a prognostic indicator, because human tumours are heterogeneous and consist of subpopulations of cells having different biologic properties. Moreover, microvessel density is determined by the intercapillary distance. During tumour regression induced by an angiogenesis inhibitor, microvessel density may decrease if capillary dropout exceeds tumour cell dropout, increase if tumor cell dropout exceeds capillary dropout, or remain the same if a parallel disappearance of capillaries and tumour cells parallel each other (Kerbel and Folkman 2002). Therefore, the detection of a decrease in microvessel density during treatment with an angiogenesis inhibitor, suggests that the agent is active. However, the absence of a decrease in microvessel density does not correspondingly suggest that the agent is ineffective.

## 5.11 Angiogenesis Is Not Necessarily Involved in Tumour Progression

Angiogenesis in human tumours is considerably less active than in a physiological condition such as the formation of granulation tissue in the reproductive organs; in fact, the endothelial cell proliferation index value is 0.15% for the human prostate or breast cancer compared to 6.7% in granulation tissue and 36% in the corpus luteum. Moreover, the microvessel densities in human lung, mammary, renal cell and colon carcinomas, glioblastoma and pituitary adenomas are lower than those in their normal counterparts (Eberhard et al. 2002; Turner et al. 2000). In lung carcinoma, for example, the microvessel density was found to be only 29% that of normal lung tissue. In glioblastoma, microvessel density was found to be 78% that of normal brain tissue. This apparent paradox is partially explained by the lower oxygen consumption rate of tumour cells (Sterinberg et al. 1997), which are also known to tolerate oxygen deprivation (Graeber et al. 1996). As a result, the intercapillary distance in tumours is greater than in their normal tissue counterparts.

# Chapter 6
# Mast Cells and Tumours

## 6.1 Mast Cells and Tumour Growth in Humans: Pro and Contra

Already in 1891, Westphal, a pupil of Paul Ehrlich, observed an accumulation of MCs in the connective tissue of various tumours in humans (Westphal 1891). This predilection of MCs was repeatedly confirmed by various authors (Unna 1896; Fromme 1906; Bonney 1908; Weill 1919; Staemmler 1921; Higuchi 1930; Quensel 1933; Michels 1938; Sylvèn 1940; Bali and Furth 1949). Accordingly to Sylvèn (1940) the process of tumour invasion is preceded by the conversion of the adjacent normal connective tissue into a gelatinous matrix rendered metachromatic by the release of granular material from tissue MCs. In slowly growing mesenchymal tumours the preliminary process of digestion of the normal tissue so far outstrips the subsequent utilization of this 'free chromotrope substance'. That the tumour is seen to be surrounded by a vidid band of diffuse metachromasia, whereas in more anaplastic growths the metachromatic zone is consumed as fast as it is formed. Several authors have noted that in slowly growing scirrhous carcinomas the relatively abundant and well differentiated stroma usually contains an appreciable MC content (Bäumer 1896; Harris 1900; Williams 1900; Clowes and Owen 1904), whereas in more anaplastic tumours the stromal MCs are few and poor in granules (Regaud and Lacassagne 1922; Higuchi 1930; Holmgren 1946).

In 1958, Lascano compared the number of MCs in various benign and malignant tumours and demonstrated that a stricking increase of MCs was detectable only in cases of pigmented nevi and of basal cell carcinoma of the skin. The number remained unchanged, or was reduced, in cases of capillary hemangioma, squamous cell carcinoma and neurilemmoma of the skin; in fibroadenoma and scirrhous carcinoma of the breast; in carcinoma of the lung, esophagus, stomach, colon, gall bladder and uterine cervix; in lymphosarcoma and malignant melanoma; and in gastric or uterine myoma (Lascano 1958).

MCs accumulate at sites of tumour growth in response to numerous chemoattractants (Conti et al. 1997). SCF is a powerful MC chemoattractant and activator (Meininger et al. 1992; Wershil et al. 1992) and it has been implicated in MC accumulation at the periphery of tumours (Zhang et al. 2000; Huang et al. 2008).

D. Ribatti, E. Crivellato, *Mast Cells and Tumours*,
DOI 10.1007/978-94-007-1469-4_6, © Springer Science+Business Media B.V. 2011

SCF overexpression in mammary tumours increases MC accumulation at local sites of tumour growth, whereas inhibition of SCF expression results in decreased MC accumulation and angiogenesis (Zhang et al. 2000). SCF-knockdown in a H22 hepatocarcinoma model led to decreased tumour growth correlated with a reduction in MC accumulation in tumour site (Huang et al. 2008). SCF-activated MCs exacerbate tumour immunosuppression in mice by releasing adenosine and increasing Treg cells (Huang et al. 2008).

MCs have a vast array of mediators, some of which have promoting, and others, inhibitory effects on malignancies (Fig. 6.1; Theoharides and Conti 2004).

The phenotypic expression of MCs is not static and its secretory pattern varies according to the microenvironment and MCs have the ability to secrete individual

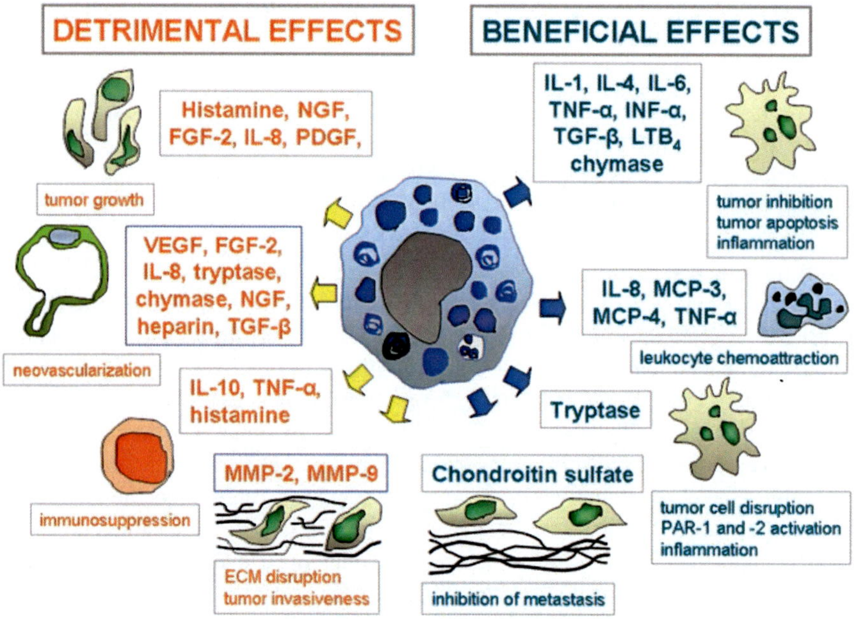


**Fig. 6.1** The dual role of mast cells in tumour fate. Mast cells may release in the tumour stroma cytokines and growth factors, such as FGF-2, NGF, PDGF, IL-10 and IL-8, which have detrimental effects to the host by stimulating tumour cell expansion. Mast cells are a major source of histamine, which can induce tumour cell proliferation through H1 receptors, while suppressing the immune system through H2 receptors. In addition, mast cells synthesize and store angiogenic factors as well as matrix metalloproteinases, which promote tumour vascularization and tumour invasiveness, respectively. Mast cells may also generate immunosuppression by releasing IL-10, histamine and TNF-α. By contrast, mast cells may promote inhibition of tumour cell growth, tumour cell apoptosis and inflammation by releasing cytokines such as IL-1, IL-4, IL-6, and TNF-α. TNF-α, in particular, is very effective in leukocyte chemoattraction. Chondroitin sulphate may inhibit tumour cell diffusion and tryptase causes both tumour cell disruption and inflammation through activation of protease-activated receptors (PAR-1 and -2). (Reproduced from D. Ribatti and E. Crivellato, Mast cell, angiogenesis and cancer. In "Mast cell biology: contemporary and emerging topics", A.M. Gilfillan and D.D. Metcalfe eds., Landes Bioscience and Springer Science, 2011)

granules or distinct mediators selectively. Secretion of cytokines from MCs could occur without classical degranulation. This has been termed "differential release", "intragranular activation", or "piecemeal degranulation" (Theoharides et al. 2007).

Dabbous et al. (1986) showed that MC degranulation is associated with disruption and lysis of the tumour extracellular matrix either directly through the action of their enzymes, or indirectly through modulation of the collagenolytic activity of fibroblasts, macrophages and tumour cells (Stack and Johnson 1994; Baram et al. 1991). Tryptase activates latent MMPs and plasminogen activator, which, in turn, degrade the extracellular matrix (Stack and Johnson 1994). MC chymase was demonstrated to cause apoptosis in different target cells (Hara et al. 1999) and to induce the accumulation of tumour associated macrophages, neutrophils and other inflammatory cells *in vivo* (He and Walls 1998).

Heparin enhances both the activity and production of collagenase *in vitro* and release plasminogen activator from endothelial cells (Sakamoto et al. 1973; Markwardt and Klocking 1977).

MCs are the major source of histamine, which mediates many symptoms of allergic reactions. Histamine could induce tumour cell proliferation through H1R identified in human malignant carcinoma, while suppressing the immune system through H2R (Niwa et al. 1991). High histamine concentration inhibited human primary melanoma-cell proliferation, presumably by acting through H1R, an action enhanced by IL-2, whereas low amounts through H2R increased proliferation (Lazar-Molnar et al. 2002). It has been reported that the incidence of metastases as well as the appearance of tumours correlates inversely with tissue histamine level (Burtin et al. 1985). MCs release TNF-α, which induces tumour cell death (Gordon and Galli 1990). Moreover, MC-derived TNF-α and interferon cooperate and enhance TRIAL gene expression, leading to apoptosis of tumour cells (Abadie et al. 2004).

MCs exert direct mitogenic effect on tumour cells. For example, FGF-2 and IL-8 are directly mitogenic to melanocytes and melanoma cells (Halaban et al. 1988; Schadendorf et al. 1993).

We have investigates the pattern of distribution of MCs in biopsy samples obtained from four different human tumours, utilizing an image analysis system and a mathematical model to make a quantitative approach to characterizing their spatial distribution (Guidolin et al. 2009). In all tumours MCs demonstrated a virtually random spatial distribution, albeit with varying densities, suggesting that MC interactions could play a minor role in the formation of the MC pattern in neoplastic tissues. The random distribution of the cells in the tissues could be accounted for by a random walk migration under the influence of cell-matrix interactions or chemotactic fields potentially generated by tumour or endothelial cells.

Tumours of MCs are extremely rare, but certainly not as rare as commonly thought. The most common tumours of MCs are those confined to the skin and manifest in childhood (Lennert and Parwaresch 1979). The group of the cutaneous mastocytosis consists of two types. The most common is designated urticaria pigmentosa and represents a disseminated involvement of the skin, while the localized type is made of solitary or multiple nodules and is called benign mastocytoma

(Lennert and Parwaresh 1979). The clinical course in cutaneous mastocytosis is benign, in many cases, skin lesions disappear during puberty (Caplan 1963).

Systemic mastocytosis is usually diagnosed in adulthood and is characterized by multiorgan involvement (with or without skin lesions) and disease persistence (Lennert and Parwaresh 1979). Indolent variants as well as aggressive variants of systemic mastocytosis have been reported (Lennert and Parwaresh 1979). Patients with systemic mastocytosis may be diagnosed with an associated clonal haematologic non-mast cell lineage disease (Travis et al. 1988) and MC-leukemia is a rare subtype of systemic mastocytosis defined by circulating MCs and a rapidly deteriorating clinical course in most cases (Travis et al. 1988).

Increased MC number has been correlated with a poor prognosis in several human tumours, such as human melanoma (Ribatti et al. 2003a), oral squamous carcinoma (Iamaroon et al. 2003) and squamous cell carcinoma of the lip (Rojas et al. 2005).

Sharma et al. (1992) reported a higher number of MCs in nodular sclerotic-type Hodgkin's lymphoma (HL) than in other types of HL and the number of MCs was higher in fibrotic areas than in cellular areas. Molin et al. (2002), Molin (2004) observed a worse prognosis for a nodular sclerosing HL exhibiting a high MC number. Fukushima et al. (2006) demonstrated an increased number of tryptase-positive and chymase-positive MCs in fibrotic areas in diffuse large B-cell lymphoma (DLCCL) lymph nodes. The greatest number of MCs among T-cell lymphomas we observed in angioimmunoblastic T-cell lymphoma (Fukushima et al. 2001). Dave et al. (2004) utilized gene array to study the relationship between prognosis and a specific gene expression profile and concluded that the length of survival among patients with follicular lymphoma correlates with the molecular features of non-malignant immune cells present within the tumour at the time of diagnosis. One of the genes observed to correlate most negatively with survival was that of microphathalmia-associated transcription factor (MITF), a transcription factor which has been found to be highly expressed in MCs and to play a critical role in the regulation of several key mast cell-specific genes (Nechushtan and Razin 2002). Tournilhac et al. (2006) demonstrated that MCs may support tumour cell expansion in Waldenstrom's macroglobulinemia through constitutive CD154-CD40 signalling. In detail, MCs expressed CD154, a potent inducer of malignant B-cell proliferation, while bone marrow lymphoplasmacytic cells functionally expressed the CD154 receptor, CD40. Moreover, the use of CD154-CD40 signal partially inhibited MC mediated bone marrow lymphoplasmacytic proliferation and/or tumour colony formation.

Nonomura et al. (2007) demonstrated that in prostate cancer MC counts were higher around cancer foci in patients with higher Gleason scores that in those with low Gleason scores. The MC number correlated with clinical stage and multivariate analysis revealed that MC count was a significant prognostic factor.

MCs are predominantly localized at the tumour periphery, at the interface with healthy tissues, rather than within the tumour. MC recruitment occurs from resident MCs migrating from neighboring healthy tissue or through *de novo* recruitment of MCs progenitors from vasculature close to the tumour site. For example, immature myeloid cells recruited to the front of developing tumours express CD34, CCR1,

MMP-2 and MMP-9, expressed by MCs at different stages of their development (Kitamura et al. 2007).

Increased MC number has been correlated also with a good prognosis in several human tumours. In this case, an anti-tumour role of MCs reflects the ability of these cells to mediate direct tumour killing.

Gomes et al. (2008) detected an increased MC density in actinic cheilitis and in squamous cell carcinoma of the lip compared to normal oral mucosa, suggesting a role for MC in the development and progression of these lesions. Similar findings were reported by Lago Costa et al. (2009). In another study, COX-2 overexpression, tryptase-positive MCs and PAR-2 were investigated in normal lip biopsies and in lip biopsies with actinic cheilitis (Rojas et al. 2009). MC-derived tryptase can induce COX-2 expression by the cleavage of PAR-2. Increased epithelial co-expression of COX-2 and PAR-2, as well as elevated subepithelial density of tryptase-positive MCs were found in actinic cheilitis compared to normal lip. COX-2 overexpression was found to be a significant predictor of actinic cheilitis, and to be correlated with both tryptase-positive MCs and PAR-2 expression. The authors concluded that tryptase may contribute to COX-2 upregulation by epithelial PAR-2 activation during early lip carcinigenesis (Rojas et al. 2009). Remarkably, the number of tryptase- and KIT-positive MC was found to be decreased in oral squamous cell carcinoma biopsies by Oliveira-Neto et al. (2007).

Aaltomaa et al. (1993) found a positive correlation between survival and increased MC number in a study of 187 breast cancer biopsies. Dabiri et al. (2004) analyzed the correlation between MC number in breast cancer and patients' prognosis in a study of 438 patients. They found a strong correlation between the presence of MCs and a favorable prognosis. Perivascular tumour-associated MCs in mammary adenocarcinoma could secrete several cytokines and proteolytic enzymes that could be detrimental to the tumour cells. For instance, IL-4, which binds to IL-4 receptors expressed by human breast carcinoma cells, could lead to apoptosis in breast cancer (Gooch et al. 1998). The histamine content of human breast cancer tissue is much higher than adjacent normal tissue and act as a local immunosuppressant (Ohno et al. 2002). Moreover, the mean level of serum tryptase in women with breast cancer is three-times higher than in healthy women (Samoszuk et al. 2003). Kankkunen et al. (1997) observed that significant increase in MC counts in breast carcinoma versus benign lesions is due to tryptase-containing MCs. It was found that in benign lesions, the number of MCs exhibiting tryptase activity was similar to that of chymase-active MCs. Malignant tumours, however, had two to three times more tryptase-containing than chymase-containing MCs, while the tryptase activity was significantly higher than in benign lesions. Moreover, in malignant lesions, tryptase-containing MCs were concentrated at the tumour edge, whereas chymase-containing MCs were not increased in this area. Breast cancer patients with metastases in the axillary nodes reveal greater number of MCs in all nodes examined compared with patients without metastasis (Thorense et al. 1982).

Welsh et al. (2005) analyzed the presence of MCs in the tumour stroma of 175 patients with surgical resected non small cell lung carcinoma and demonstrated that both macrophage and MC infiltration of the tumour islets was associated with a marked increase in 5-years survival, independently of other favorable prognostic factors including stage.

Chan et al. (2005) studied samples of ovarian cancer from 44 patients and demonstrated that tumours with higher microvascular density had a higher mean survival compared with low MC density or low microvessel density. Cinel et al. (2009) showed a significant correlation between high MC density and the presence of myometrial invasion in endometrial carcinomas. Tumour infiltrating tryptase-positive MCs, after IL-2 preoperative induction therapy, predict improved clinical outcome in patients with malignant pleural mesothelioma, and highlight the critical role of the local inflammatory response in mesothelioma cancer progression (Alì et al. 2009).

# Chapter 7
# Mast Cells and Tumour Angiogenesis

Inflammatory cells regulate endothelial cell functions related to physiological angiogenesis as well as inflammatory and tumour-associated angiogenesis. It was Rudolf Virchow in 1863, who critically recognized the presence of inflammatory cells infiltrating neoplastic tissues and first established a causative connection between the "lymphoreticular infiltrate" at sites of chronic inflammation and cancer.

In neoplastic tissues, inflammatory cells act in concert with tumour cells, stromal cells and endothelial cells to create a microenvironment, which is critical for the surviving, development and diffusion of the neoplastic mass. These synergies may represent important mechanisms for tumour development and metastasis by providing efficient vascular supply and easy pathway to escape. Indeed, the most aggressive human cancers, such as malignant melanoma, breast carcinoma, gastric cancer and colorectal adenocarcinoma, are associated with a dramatic host response composed of various inflammatory cells, especially macrophages and MCs. These observations suggest a role for MCs in the shift between inflammation and cancer (Ribatti and Crivellato 2009).

## 7.1 Mast Cell-Induced Angiogenesis Studied by Means of *In Vitro* and *In Vivo* Assays

Rat peritoneal MCs migrate in response to conditioned medium from a number of tumorigenic cell lines, but not primary cells or non-tumorigenic cell lines, suggesting the existence of tumour-intrinsic factors in MC recruitment (Poole and Zetter 1983). These authors proposed that the chemoattraction of MCs by tumour-derived peptides may be an important early event in tumour neovascularization.

Kessler et al. (1976) demonstrated that tumour angiogenesis factor (TAF) elicited a vasoproliferative response when implanted upon the CAM of the chick embryo. This response was first observed stereomicroscopically 2–3 days after implantation and a 40-fold increase in MC density was observed in the vicinity of the implants by 24 hours.

D. Ribatti, E. Crivellato, *Mast Cells and Tumours*,

DOI 10.1007/978-94-007-1469-4_7, © Springer Science+Business Media B.V. 2011

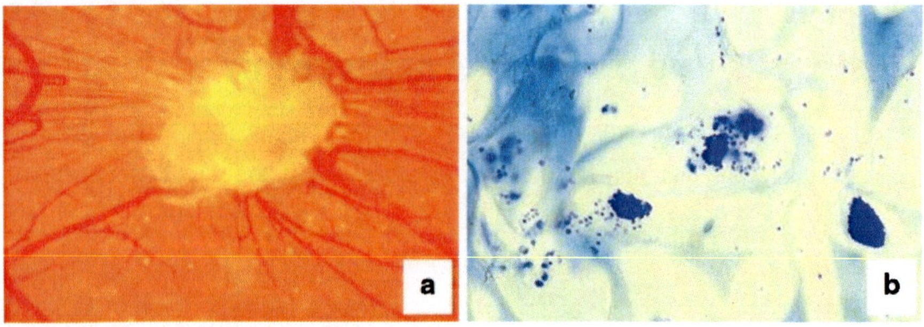


**Fig. 7.1** A mast cell suspension has been delivered on the top of the chick embryo chorioallantoic membrane on day 8 of incubation using a gelatin sponge implant. Macroscopic observation on day 12 shows the gelatin sponge surrounded by numerous allantoic vessels that develop radially towards the implant in a "spoked-wheel" pattern (**a**). The histological analysis shows among the sponge trabeculae metachromatic mast cells and their secretory granules (**b**). (Reproduced from Ribatti and Crivellato 2009)

Isolated rat MCs and their secretory granules, but not degranulated MCs, induce an angiogenic response in the CAM assay (Fig. 7.1) (Ribatti et al. 2001a). Addition of anti-FGF-2 or anti-VEGF antibodies reduced the angiogenic response of both MCs and their secretory granules by 50 and 30% respectively. These data support the evidence that the angiogenic properties of MCs depend on the angiogenic molecules contained in their secretory granules, and indicate that FGF-2 and VEGF are the angiogenic cytokines primarily and perhaps synergistically responsible for this vasoproliferative activity. More recently, Detoraki et al. (2009) demonstrated that primary human lung MCs are angiogenic in the CAM assay and this effect is inhibited by an antibody anti-VEGF-A. MCs are angiogenic *in vivo* in other assays, such as the rat mesentery assay (Norrby et al. 1986) and the limb ischemic reperfusion assay (Heissig et al. 2005).

## 7.2 Mast Cells-Induced Angiogenesis Studied by Means of Experimental Carcinogenesis

Data on MC function in developing tumours has resulted form mouse models of cancer and it is in the context of experimental carcinogenesis that MC contribution to tumour angiogenesis is most evident.

Holmgren and Wohlfart (1947) demonstrated that in chemically induced fibrosarcomas MCs were scattered diffusely among the malignant fibroblasts and through the vascularized tumours perivascular MCs were comparatively rare. Bali and Furth (1949) showed in transplantable splenic tumour that MCs were numerous in the capsule of the tumour and were abundant in the connective stroma and in the adventitia of vessels.

Experimentally induced tumors display MC accumulation close to the tumour cells before the onset of angiogenesis (Kessler et al. 1976). Likewise, an increase in

MC number has been observed in tumour invasion around rat mammary adenocarcinoma (Dabbous et al. 1986). Early studies in experimentally induced epidermoid carcinoma in hamster buccal pouches by repeated topical application of dimethylbenzanthracene demonstrated sequential MC migration towards progressive mucosal dysplasia and subsequent development of squamous cell carcinoma (Flynn et al. 1991).

Starkey et al. (1988) investigated the role of host MCs in tumour-associated angiogenesis by comparing the angiogenic response of genetically MC-deficient $W/W^v$ mice and MC-sufficient +/+ litter mates to subcutaneously growing B16-BL6 tumours. The response was slower and initially less intense in $W/W^v$. Fewer $W/W^v$ than +/+ mice developed spontaneous lung metastases. Bone marrow repair of the MCs deficiency restored the incidence of haematogeneous metastases to approach that of +/+ mice. These results demonstrate a role for MCs *in vivo* during tumour angiogenesis and also suggest a role for host MCs in haematogeneous metastases.

The introduction of novel experimental procedures to induce carcinogenesis in laboratory animals has contributed to increase our understanding of the role of MCs in early tumour angiogenesis. Development of squamous cell carcinoma in a HPV (human papilloma virus) 16-infected transgenic mouse model of epithelial carcinogenesis provided elegant experimental clues in favour of an early participation of MCs and MC-related angiogenesis in tumour growth (Coussens et al. 1999). Neoplastic progression in this model was accompanied by upregulation of genes encoding several proangiogenic growth factors, implicating these molecules in tumour angiogenesis in the skin. Infiltration by MCs and activation of the MMP-9 coincided with the angiogenic switch in premalignant lesions through the release of bound proangiogenic compounds from the extracellular matrix. MCs were shown to infiltrate hyperplasias, dysplasias and the invasive front of carcinomas, but not the core of solid tumours. MCs were seen to degranulate in close apposition to capillaries and epithelial basement membranes, releasing the MC-specific tryptase and chymase. Addition of chymase alone was sufficient to stimulate an angiogenic phenotype when co-incubated with a hyperplastic skin sample *in vitro*, while tryptase was involved in tissue remodelling and α-1 collagen production. Remarkably, premalignant angiogenesis was abated in a MC-deficient HPV16 transgenic mouse indicating that neoplastic progression in this model involved participation of MCs infiltrating the skin.

It is generally accepted that most colon cancers develop from adenomatous polyps and, in a recent experimental investigation in mice, it has been demonstrated that, from the onset, polyps are infiltrated with proinflammatory MCs and their precursors (Gounaris et al. 2007). These authors used APCΔ468 mice and a transgenic mouse overexpressing an inducible, stabilized β catenin in intestinal enterocytes, two independent hereditary models of polyposis, and demonstrated that in both the experimental models developing polyps were rapidly infiltrated by $CD34^+$ MCs in a T and B cell-independent manner. Depletion of MCs either pharmacologically with anti-TNF-α antibodies, or through the generation of chimeric mice with genetic lesions in MC development leads to a profound remission of existing polyps, suggesting that MCs are an essential haematopoietic component for preneoplastic polyp development as well as a possible target for therapeutic intervention.

A similar result has been obtained with a different experimental approach in mice. Myc is a highly pleiotropic transcription factor whose aberrant activation both initiates and maintains many aspects of the neoplastic phenotype, including both cell-intrinsic proliferative programs and extracellular programs that instruct changes in the tumour microenvironment such as stromal remodelling, angiogenesis and invasion. In a pancreatic β-cell tumour model, activation of Myc *in vivo* has been shown to trigger rapid recruitment of MCs to the tumour site, recruitment that is absolutely required for macroscopic tumour expansion (Soucek et al. 2007). Macrophages and neutrophils are also recruited to the islets, although later and after the onset of tumour angiogenesis. In addition, treatment of established β-cell tumours with cromolyn, an inhibitor of MC degranulation, rapidly triggers hypoxia and cell death of tumour and endothelial cells. Similarly, activation of Myc in β-cells of genetically MC-deprived mice does not elicit any measurable β-cell or islet expansion. Again, inhibitors of MC function may therefore prove therapeutically useful in restraining expansion and survival of pancreatic and other cancers.

Multiple myeloma in humans is frequently associated with MCs infiltration and neovascularization, which correlate directly with disease severity (Ribatti et al. 1999b). Ang-1 appears as a crucial factor in promoting multiple myeloma cell growth by stimulating angiogenesis. Experimental evidence indicates that Ang-1 secreted by primary murine MCs promote marked neovascularization in implanted Matrigel plugs (Nakayama et al. 2004). Primary MCs accelerate tumour growth by established plasmacytoma cell lines and the use of Ang-1-neutralizing antibodies reduces significantly the growth of plasmacytomas containing MCs. These results demonstrate that MC-derived Ang-1 promotes growth of plasmacytomas by stimulating neovascularization and provide further evidence supporting a casual relationship between inflammation and tumour growth.

Huang et al. (2008) demonstrated that tumour-derived SCF recruits MCs to the tumour environment and also activates them. Using a SCF-knockdown H22 tumour model, tumour growth was decreased and this correlated with a reduction in MC accumulation in the developing tumour. SCF-stimulated MC released active MMP-9 into the local environment, disrupting extracellular matrix and releasing further matrix-bound SCF, acting as a positive feed-back loop on MC activation within the tumour. Activation by high levels of SCF *in vitro* led to upregulation of proinflammatory genes, such as IL-6, TNF-α and VEGF, which could be inhibited by antibodies against the SCF receptor, KIT.

Hart et al. (2000), using the $W/W^{v}$ mice demonstrated a direct correlation between MC density in the dermis and susceptibility to ultraviolet B-induced systemic immunosuppression. These mice which are homozygous for the *W* (white spotting) mutation, and therefore severely MC deficient are unresponsive to ultraviolet-induced immunosuppression unless first injected with MC precursors at the irradiated site (Hart et al. 2000). The *W* locus encoded the KIT tyrosine kinase receptor that binds SCF (Church and Levi-Schaffer 1997). The MC products involved in ultraviolet-induced immunosuppression are believed to be TNF-α and histamine (Hart et al. 2001).

## 7.3 Angiogenenic Factors Stored in Mast Cells

MCs synthesize several proangiogenic molecules (Table 7.1 and Fig. 7.2). In 1980, Azizkhan et al. reported that heparin derived from MCs could stimulate the migration, but not the proliferation of bovine capillary endothelial cells (Azizkhan et al. 1980), while Fraser and Simpson reported that commercial heparin could stimulate proliferation of human umbilical vein endothelial cells as well as angiogenesis in the CAM assay (Fraser and Simpsom 1980).

In 1982, Taylor and Folkman showed that heparin could enhance angiogenesis induced by tumour extracts implanted in the chick embryo. New capillaries appeared 24 h earlier with heparin than without it. Protamine, a heparin antagonist, prevented heparin from enhancing tumour angiogenesis and was able to inhibit tumour angiogenesis (Taylor and Folkman 1982). Protamine also suppressed MC-mediated angiogenesis (Jakobson et al. 1990).

Heparin stimulates endothelial cell proliferation and migration *in vitro* (Thorton et al. 1983; Alessandri et al. 1984). *In vivo*, however, it stimulates (Fig. 7.3) (Ribatti et al. 1987; Norrby and Sorbo 1992; Norrby 1993), inhibits (Jakobson and Hahnenberger 1991; Wilks et al. 1991; Norrby 1993) or has no effect (Castellot et al. 1982; Taylor and Folkman 1992). These differences seem to be related to its molecular size (high-molecular-weight heparin has been shown to stimulate angiogenesis, whereas low-molecular-weight inhibits it) and degree of sulphation. Heparin acts as a soluble form of the low-affinity FGF-2 receptor (Folkman and Shing 1992), which displaces FGF-2 in the biologically active form, and allows its rapid interaction with endothelial cells (Yayou et al. 1991). Release of FGF-2 from its storage in the extracellular matrix elicits a localized proliferation of endothelial cells and the release may be brought about by heparin or by heparin-degrading enzymes. It has been proposed that basement membranes may act as a storage depot for FGF-2

**Table 7.1** Angiogenic factors stored in mast cells

| |
|---|
| Adrenomedullin |
| Chymase |
| FGF-2 |
| Heparin |
| Histamine |
| IL-8 |
| MMP-2 |
| MMP-9 |
| NGF |
| TGF-β |
| Tryptase |
| TNF-α |
| VEGF |

*FGF-2* fibroblast growth factor-2, *IL-8* interleukin-8, *MMP-2* matrix metalloproteinase-2, *MMP-9* matrix metalloproteinase-9, *NGF* nerve growth factor, *TGF-β* transforming growth factor beta, *TNF-α* tumor necrosis factor alpha, *VEGF* vascular endothelial growth factor

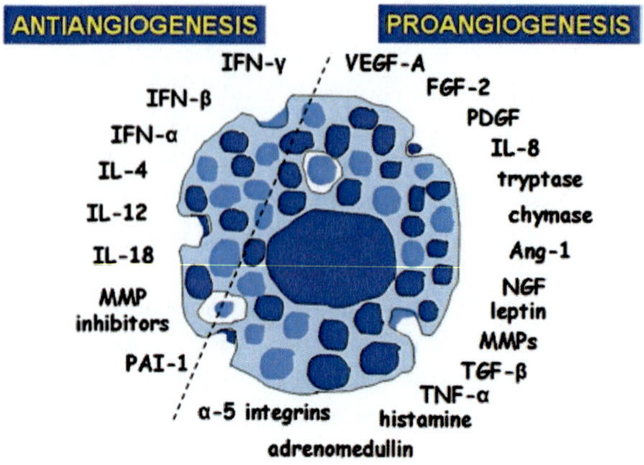


**Fig. 7.2** Mast cells synthesize and release a number of products which exert either stimulatory or inhibitory effect on blood vessel formation. The list of proangiogenic factors (on the *right*) exceeds the record of antiangiogenic mediators (on the *left*) and, in a global sense, mast cells are regarded as tissue-homing inflammatory cells playing a crucial role in the amplification of the angiogenic response. This functional activity is of the uppermost importance in tumour setting whereby mast cells may favour tumour expansion by stimulating the network of vascular supply. Among others, VEGF-A and the serine protease tryptase are the most powerful angiogenic products secreted by mast cells

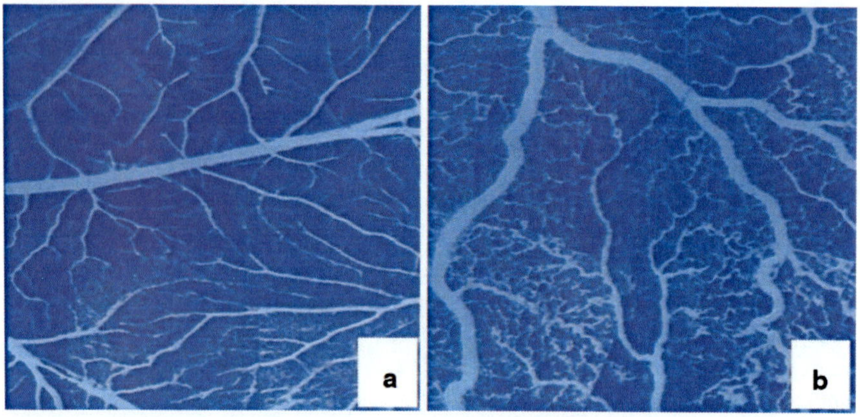


**Fig. 7.3** Photographic reconstruction of a portion of chick embryo chorioallantoic membrane treated with heparin (**b**) compared with a control (**a**). Note that in the control the vessels run straight and interdigitate regularly, whereas in heparin-treated specimen, the vessel course is irregular or sinuous, their branching is frequent and vascular loops are numerous. (Reproduced from D. Ribatti, The chick embryo chorioallantoic membrane in the study of angiogenesis and metastasis. Springer Science+Business Media B.V., 2010)

*in vivo* and that the release of FGF-2 from these stores might be mediated by production of heparin-degrading enzymes produced by tumour cells and inflammatory cells found in tumours (Nakajima et al. 1984). Heparanase is released by MCs (Bashkin et al. 1990) and can release active heparan sulfate-bound FGF-2 from the extracellular matrix of endothelial cells (Vlodavsky et al. 1992).

Samoszuk and Corvin (2003a) demonstrated in a mammary adenocarcinoma tumour model in mice an increased tumour growth when MC-derived heparin was inhibited.

Histamine has been reported to be angiogenic *in vivo* (Fraser and Simpson 1980; Zauberman et al. 1969) and mitogenic for capillary endothelial cells *in vitro* (Roche 1986; Marks et al. 1986). Histamine has an angiogenic effect through both H1R and H2R (Sorbo et al. 1994). It may also increase the permeability of newly formed microvessels during tumour angiogenesis, and hence increase the leakage of plasma proteins and deposition of fibrin. Degradation products of fibrin are angiogenic *in vivo* (Thompson et al. 1995).

MCs synthesize and store large amounts of MMP-2, MMP-9 and serine-proteinases of two subclasses: tryptase and chymase (Fang et al. 1999; Di Girolamo and Wakefiled 2000; Vincent et al. 2000). Given the ability of MMP-2 and MMP-9 to degrade type IV, V, VII and X collagens, as well as fibronectin (Raza and Cornelius 2000), namely the major components of the interstitial stroma and subendothelial basement membrane, the findings suggest that MCs may contribute to the progression from *in situ* to invasive and metastatic solid tumours, characterized by an enhanced angiogenesis and secretion of proteolytic enzymes (Raza and Cornelius 2000).

Tryptase and chymase are involved in angiogenesis after their release from activated MC granules. Their proteolytic activities degrade extracellular matrix components or release matrix-associated growth factors (Taipale et al. 1995), and they also act indirectly by activating latent MMPs (Gruber et al. 1989) and plasminogen activators (Stack and Johnson 1994). Blair et al. (1997) have demonstrated the angiogenic potential of tryptase *in vitro* and its important role in neovascularization. Tryptase added to microvascular endothelial cells cultured on Matrigel caused a pronounced increase in capillary growth, and this was suppressed by specific tryptase inhibitors. Moreover, tryptase directly induced endothelial cell proliferation in a dose-dependent fashion. Recently, we have demonstrated that exogenous tryptase and chymase stimulate angiogenesis in the CAM assay (Ribatti et al. 2011). A colocalization of tryptase and leptin and of tryptase and cathepsin G has been demonstrated in MC secretory granules (Figs. 7.4 and 7.5) (Ribatti et al. 2007a, 2009).

Muramatsu et al. (2000a, b) used the hamster sponge-implant model to show that angiogenesis is induced by angiotensin II and inhibited by chymase inhibitors, suggesting that MC-derived chymase is an important mediator of MC-dependent angiogenesis.

Adrenomedullin protect endothelial cells from apoptosis, promotes tumour angiogenesis, and affects vascular tone and permeability (Ribatti et al. 2005a). Zidaire et al. (2006) demonstrated that adrenomedullin induces histamine release from rat and human MCs and that this is chemotactic for human MCs and stimulates mRNA expression of VEGF, FGF-2 and MCP-1. Moreover, they identified

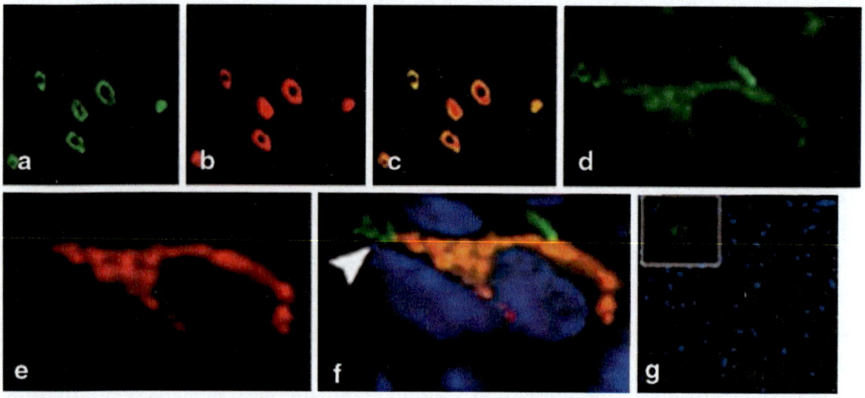


**Fig. 7.4** **a–c** Confocal scanner dual immunofluorescence of leptin (*geen*) and tryptase (*red*) and for both proteins (*orange*) in a uterine leiomyomas section, showing mast cells positive for both. **d–f** Confocal scanner dual immunofluorescence of leptin (*green*, in **d**) and tryptase (*red*, in **e**) and for both proteins (*orange*, in **f**) in a uterine leiomyoma section, showing mast cell granules positive to leptin and tryptase in the cytoplasm and around the nucleus and positive to leptin in a peripheral cytoplasmic area (*arrowhead* in **f**). **g** Absorption of the leptin antibody with 30 μmol/L leptin abolished all immunostaining. (Reproduced from Ribatti et al. 2007a)

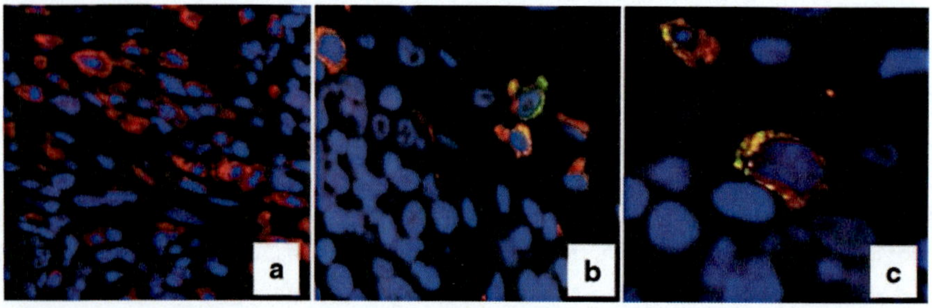


**Fig. 7.5** Dual immunofluorescence for tryptase (*red*, in **a**) and cathepsin-G (*green*, in **b**) and for both proteins (*orange*, in **c**) in a biopsy specimen of human cutaneous mastocytosis. (Reproduced from Ribatti et al. 2009)

adernomedullin-producing MCs in tumour infiltrates of human breast and lung cancer patients. Finally, a human MC line induced angiogenesis *in vivo* and a neutralizing anti-adrenomedullin monoclonal antibody blocked this ability. Tsuruda et al. (2006) demonstrated an intense immunoreactivity for adrenomedullin in MCs of the outer media and adventitia of human abdominal aortic aneurysms and Belloni et al. (2006) provided evidence that a subset of cardiac MCs are able to synthesize and store adrenomedullin.

MCs release several polypeptide growth factors, including FGF-2, VEGF, TGF-β, TNF-α and IL-8. These cytokines are involved both in normal as well as tumour-associated angiogenesis. The spectrum of cytokines expressed appears to vary depending on the maturity state of the MCs and of the tissue of residence. MC-derived TNF-α can stimulate endothelial cell collagenase production either directly or indirectly by inducing synthesis of FGF-2 which act in an autocrine manner to enhance induction of collagenase (Okamura et al. 1991). Qu et al. (1998) demonstrated that FGF-2 is localized in the cytoplasmic and extruded granules of MCs in several human tissues. Grutzkau et al. (1998) demonstrated the expression of VEGF in the human MC line HMC-1 and in human skin MCs. Detoraki et al. (2009) reported that human primary lung MCs constitutively produce VEGF-A, VEGF-B, and the lymphangiogenic VEGF-C and VEGF-D. Moreover, human lung MCs express VEGFR-1 and VEGFR-2 and different VEGFs exert chemotactic effects on lung MCs by engaging both receptors. SCF increases MC release of both VEGF and FGF-2 (Zhang et al. 2000).

Kim et al. (2010) demonstrated that the proinflammatory cytokine IL-1β induced the synthesis of the angiogenic factor IL-8 in human MCs via the $LTB_4$ receptor BLT-2. Moreover, the conditioned media collected from IL-1β-treated human MC line HMC-1 had significantly enhanced angiogenic activity that could be attenuated by either small interfering RNA knockdown of BLT-2 or treatment with neutralizing antibody anti-IL-8.

NGF, also contained in MC secretory granules, induces endothelial cell proliferation *in vitro* and angiogenesis *in vivo* in the chick embryo CAM assay (Cantarella et al. 2002; Emanueli et al. 2002). Moreover, in some tumours of neural and non-neural origin, NGF may stimulate at the same time the proliferation of tumour and endothelial cells (Nico et al. 2008a).

## 7.4 Mast Cells and Angiogenesis in Human Solid Tumours (Table 7.2)

Tryptase-positive MCs increased in number and vascularization increased in a linear fashion from dysplasia to invasive cancer of the uterine cervix (Benitez-Bribiesca et al. 2001) and in uterine leiomyomas (Ribatti et al. 2007a). We have demonstrated

**Table 7.2** Human solid tumors in which it has been demonstrated a positive correlation between mast cell infiltration and angiogenesis

| |
|---|
| Breast cancer |
| Colorectal cancer |
| Endometrial carcinoma |
| Esophageal squamous cell carcinoma |
| Gastric cancer |
| Hepatocellular carcinoma |
| Laryngeal carcinoma |
| Melanoma |
| Non-small cell lung carcinoma |
| Oral squamous cell carcinoma |
| Pancreatic adenocarcinoma |
| Prostate cancer |
| Renal cell carcinoma |
| Uterine cervix carcinoma |

that angiogenesis in human endometrial carcinoma is highly correlated with MC tryptase-positive cell counts and that these parameters increase in agreement with tumour progression (Ribatti et al. 2005b). Wilk et al. (2010) demonstrated by immunohistochemistry that microvessel density and density of tryptase-positive MCs increased in cervical tumours from normal samples through intraepithelial lesions to invasive carcinoma. Moreover, they demonstrated that the density of chymase-positive MCs is significantly higher in invasive carcinoma than in normal samples. Utrera-Barillas et al. (2010) evaluated the relationship between MC and macrophage density with blood and lymphatic vessels in different stages of carcinoma of the uterine cervix and demonstrated a correlation between MCs and blood vessels in both stage 2 of cervical intraepithelial neoplasia and carcinoma *in situ*.

A positive correlation between MC density and microvascular density has been established in esophageal squamous cell carcinoma (Tomita et al. 2001), gastric cancer (Fig. 7.6) (Ribatti et al. 2010) primary colorectal cancer (Acikalin et al. 2005; Gulobova and Vlaykova 2009), hepatocellular carcinoma (Peng et al. 2005), pancreatic adenocarcinoma (Esposito et al. 2004), renal cell carcinoma (Tuna et al. 2006), melanoma (Fig. 7.7; Ribatti et al. 2003b) and oral sqamous cell carcinoma (Sharma et al. 2010). Melillo et al. (2010) demonstrated that the density of tryptase-positive MCs in human papillary thyroid carcinomas is higher as compared to normal thyroid tissue. Moreover, they showed that different MC lines induce thyroid

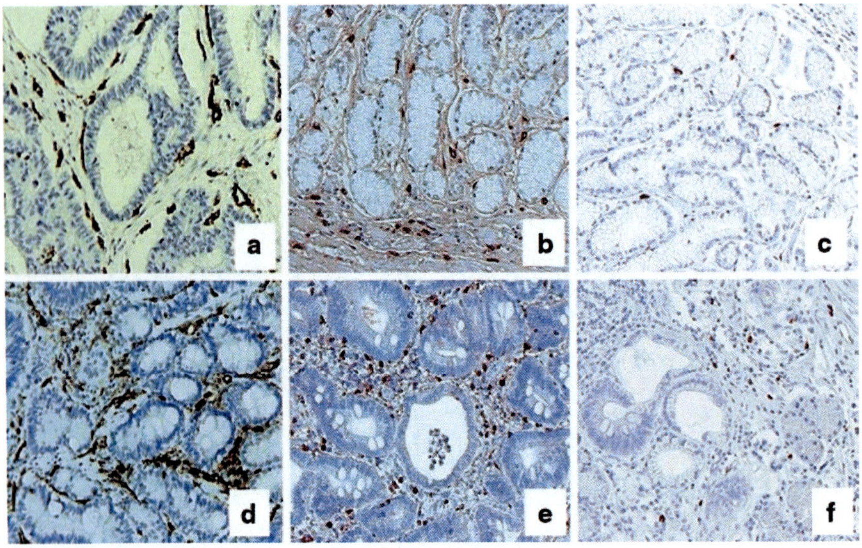


**Fig. 7.6** Immunohistochemical staining for CD31, tryptase and chymase in stage II (**a–c**) and stage IV (**d–f**) human gastric cancer. In (**a**, **d**) endothelial cells immunoreactive for CD31; in (**b**, **e**) tryptase-positive mast cells; in (**c**, **f**) chymase-positive mast cells. Blood vessels and mast cells are distributed around the gastric glands. The number of blood vessels and mast cells is higher in stage IV as compared to stage II bioptic specimens and chymase-positive are lower as compared to tryptase-positive mast cells. (Reproduced from Ribatti et al. 2010)

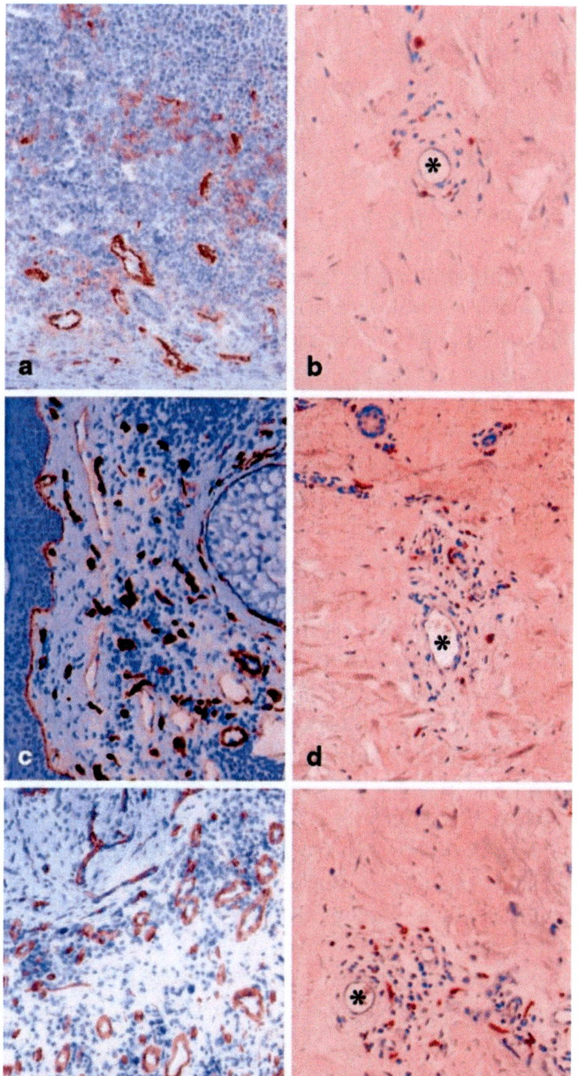


**Fig. 7.7** Adjacent sections of common nevi (**a**, **b**), nevi with varying degrees of melanocytic atypia (ADMA) (**c**, **d**), and metastatic (**e**, **f**) melanoma stained with CD31 (**a**, **c**, **e**) for microvessels and with tryptase for mast cells (**b**, **d**, **f**). Note the progressive increase of microvessels and mast cells located around blood capillaries (*asterisks*) from common nevi to nevi with ADMA and from latter to metastatic melanoma. (Reproduced from Ribatti et al. 2003)

cancer cell invasive ability, survival and DNA synthesis *in vitro* and that xenografts of thyroid carcinoma cells recruit MCs injected into the tail vein of mice. Finally, co-injection of human MCs accelerate the growth of thyroid cancer cells in athymic mice and this effect mediated by an increased tumour vascularization and proliferation, is reverted by treating mice with sodium cromoglycate.

An association of VEGF and MCs with angiogenesis has been demonstrated in laryngeal carcinoma (Sawatsubashi et al. 2000), in non-small cell lung carcinoma, where most intratumoral MCs express VEGF (Imada et al. 2000; Takanami et al. 2000; Tomita et al. 2000), and in melanoma, where MCs express both VEGF (Toth et al. 2000) and FGF-2 (Ribatti et al. 2003b). Lastly, a prognostic significance has been attributed to MCs and microvascular density in squamous cell cancer of the oesophagus (Elpek et al. 2001) and in melanoma (Ribatti et al. 2003a).

Imada et al. (2000) studied 85 cases of stage I non-small cell lung carcinoma and demonstrated a higher number of tryptase-positive MCs compared to cases of squamous cell carcinoma. A high correlation was observed between intratumoral MC counts and microvessel counts and double staining showed that most intratumoral MCs express VEGF. These data have been confirmed by Ibaraki et al. (2005). Carlini et al. (2010) observed in non-small cell lung cancer a higher MC and microvascular density in the peritumoral zone than in the intratumoral zone. Moreover, a positive correlation between tryptase/chymase positive MCs and microvascular density was observed in the intratumoral zone confirming the involvement of MCs in the angiogenic process.

In melanoma, MC accumulation around the margin of tumours has been observed to peak just as the tumour acquire the angiogenic phenotype and peritumoral MC counts correlated strongly with microvascular density, melanoma progression and prognosis (Toth et al. 2000; Ribatti et al. 2003b). Furthermore, in melanoma, MCs were closer to each other and to the vessels (Guidolin et al. 2006). This close association between MCs and the endothelium might indicate that MCs are involved in the maintenance reaction necessary for the long lasting functional integrity of the endothelium.

We have demonstrated that angiogenesis increases in parallel to the number of tryptase-positive MCs in lymph nodes from patients with breast cancer and that their values are significantly higher in lymph nodes with micrometastases as compared with those without metastasis (Ribatti et al. 2007d).

In prostate cancer, Johansson et al. (2010) have shown that intratumoral and peritumoral MCs have antagonists properties. In fact, intratumoral MCs negatively regulate angiogenesis and tumour growth, whereas peritumoral MCs enhance angiogenesis and tumour growth and express high levels of FGF-2. They demonstrated in a rat orthotopic model that after the injection of tumour cells, FGF-2 expressing MCs are recruited to the surrounding tumour microenvironment, but not within the tumour. When the tumour is small, tumour growth and angiogenesis can be modulated by the use of inhibitors, i.e. sodium cromoglycate, and stimulators, i.e. compound 48/80, of MC degranulation. Moreover, castration therapy that temporarily inhibits tumour growth and angiogenesis, upregulates MC-attracting chemokines, such as TNF-α. Finally, Johansson et al. (2010) demonstrated in a cohort of untreated prostate cancer patients with a long follow-up that the accumulation of intratumoral MCs serves as a protective role and relates to a favorable prognosis. On the contrary, MCs in peritumoral tissue relate to a poor prognosis.

## 7.5 Mast Cells and Angiogenesis in Human Haematological Tumours (Table 7.3)

Angiogenesis in benign lymphadenopathies and B-cell non-Hodgkin's lymphomas (B-NHL), measured as microvessels counts, is correlated with the total and MC tryptase-positive counts, and both increase in step with the increase in Working Formulation malignancy grades (Ribatti et al. 1998, 2000). In NHL, the cellular expression of VEGF and FGF-2 as well as MC and vessel counts were assessed (Fukushima et al. 2001). The number of MCs was greater in T-cell lymphomas than in B-cell lymphomas and in all NHL a significant correlation was found between vessel count and number of MCs and between vessel count and number of VEGF-expressing cells. Double fluorescence staining of VEGF mRNA and MC tryptase revealed that MCs expressed VEGF mRNA. These data suggest that MCs intervene in angiogenesis in these lesions by expressing VEGF. Glimelius et al. (2005) evaluated the relationship between the number of MCs and the microvessel count in tissue samples from HL-involved lymph nodes by immunohistochemistry and did not demonstrate any correlation between high microvessel count and the number of MCs.

Taskinen et al. (2008) have shown that accumulation of MCs in follicular lymphoma correlates with unfavourable prognosis after immunochemotherapy. They have further demonstrated that tumour vascularity is associated with MC content and outcome in follicular lymphoma patients treated with immunochemotherapy (Taskinen et al. 2010). Tripodo et al. (2010) have shown that MCs and IL-17 producing (Th17) T cells contribute to the composition of the lymphoma-associated microenvironment of angioimmunoblastic T-cell lymphoma.

Bone marrow angiogenesis, evaluated as microvessel area, and MC counts are highly correlated in patients with inactive and active multiple myeloma and in those with monoclonal gammopathies of undetermined significance. Both parameters increase simultaneously in active multiple myeloma (Ribatti et al. 1999b). More recently, we have demonstrated by means of electron and confocal microscopy that MCs participate in the formation of the vessel wall in the bone marrow of patients with multiple myeloma (Fig. 7.8; Nico et al. 2008b).

In both B-NHL and multiple myeloma (MM), MCs rest near or around blood or lymphatic capillaries. Their ultrastructural picture includes a typical morphological semilunar feature, or piecemeal partial degranulation of their secretory granules, unlike the IgE-mediated massive degranulation that occurs during immediate hypersensitivity reactions (Ribatti et al. 1998, 1999b; Crivellato et al. 2002a, 2003c). This morphology is typical of the slow degranulation that takes place in delayed

**Table 7.3** Human haematological tumours in which it has been demonstrated a positive correlation between mast cell infiltration and angiogenesis

| |
|---|
| B-cell chronic lymphocytic leukemia |
| B-cell non Hodgkin's lymphoma |
| Follicular lymphoma |
| Multiple myeloma |
| Myelodysplastic syndrome |

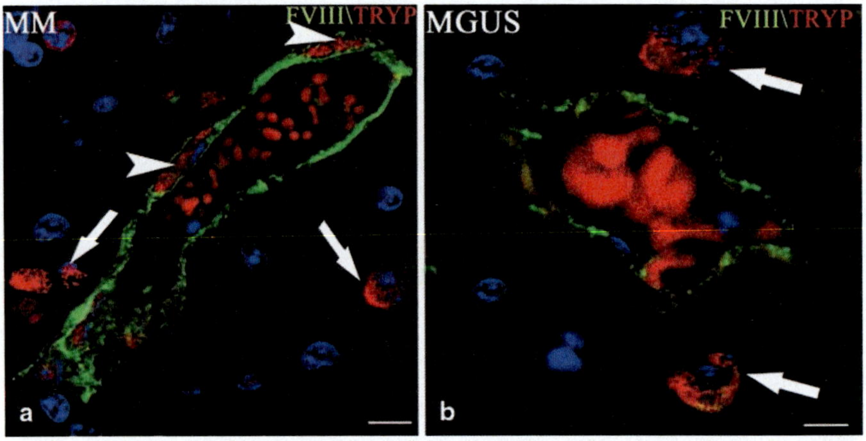


**Fig. 7.8** Double FVIII-RA (*green*) and tryptase (*red*) confocal laser microscopy from multiple myeloma (**a**) and monoclonal gammapthies of undetermined significance (**b**) bone marrow bioptic specimens. In (**a**) a multiple myeloma vessel is lined by both endothelial cells positive to FVIII-RA and by mast cells positive to tryptase (*arrowhead*). Mast cells containing tryptase-positive granules (*arrow*) are also recognizable on the abluminal side of the vessel. In (**b**) a monoclonal gammapthies of undetermined significance vessel is lined only by endothelial cells positive to FVIII-RA and is surrounded by tryptase-positive mast cells (*arrow*). (Reproduced from Nico et al. 2008b)

hypersensitivity reactions and chronic inflammation (Kops et al. 1984). Semilunar appearance may reflect slow but progressive release of angiogenic factors favouring chronic and progressive stimulation of MC degranulation.

Bone marrow samples of patients with myelodysplastic syndromes display a high correlation between microvessel counts and both total and tryptase-positive MCs, and that both parameters increase simultaneously with tumour progression (Ribatti et al. 2002).

There is also a correlation between the extent of angiogenesis and the number of tryptase-positive MCs in patients with early B-cell chronic lymphocytic leukemia and tryptase-positive MCs predict their clinical outcome (Molica et al. 2003; Ribatti et al. 2003c). We have also demonstrated that the reduction in the extent of bone marrow angiogenesis observed in B-cell chronic lymphocytic leukemia after sequential therapy with low doses of subcutaneous alemtzumab after a clinical response to fludarabine-induction therapy, was associated to a reduction in the number of bone marrow MCs (Molica et al. 2007).

# Chapter 8
# Mast Cells and Tumour Lymphangiogenesis

The lymphatic vascular system is essential for a series of important physiological and pathophysiological functions, such as lipid absorption, fluid homeostasis, and immune surveillance. Recent work has also unraveled additional functional roles of the lymphatic vasculature in fat metabolism, obesity, inflammation, and the regulation of salt storage in hypertension (Maby-El Hajjami and Petrova 2008; Wang and Oliver 2010). The lymphatic vasculature is also crucial in promoting tumour cell escape and dissemination of neoplastic cells to local lymph nodes.

It has been demonstrated that MCs not only stimulate angiogenesis but they also promote lymphangiogenesis. Indeed, recent data indicate that human MCs are both a source and a target of angiogenic and lymphangiogenic factors and therefore might play a role in inflammatory and neoplastic angiogenesis through the expression of several forms of VEGFs and their receptors (Detoraki et al. 2009). Angiogenesis and lymphangiogenesis, indeed, are important aspects of tumour growth and inflammatory diseases. Both rely on the production of different forms of VEGFs, which are the most potent proangiogenic mediators. The VEGF family includes VEGF-A, VEGF-B, VEGF-C, VEGF-D and PIGF. VEGF-A and VEGF-B are key regulators of blood vessel growth, whereas VEGF-C and VEGF-D regulate primarily lymphangiogenesis (Alitalo et al. 2005; Ferrara 2007). Indeed, endothelial receptor tyrosine kinase VEGFR-3 and its ligand VEGF-C are required for the survival, maintenance, and migration of lymphatic vessels during embryonic and post-embryonic development. In addition to VEGF-C, VEGF-D also enhances tyrosine phosphorylation of VEGFR-3 (Wang et al. 2001). Several other growth factors with lymphangiogenic function have been proposed, including FGF-2, PDGF, IGFs, hepatocyte growth factor (HGF) and growth hormone (GH) (Kubo et al. 2002; Cao et al. 2004; Chang et al. 2004; Shin et al. 2006).

Endostatin, a proteolytic fragment of collagen XVIII, is a potent inhibitor of lymphoangiogenesis and tumour growth. In transgenic J4 mice, which overexpress endostatin in their keratinocytes, carcinogen-induced skin squamous cell carcinomas were less aggressive and more often well differentiated than those in the control mice, indicating that endostatin regulates the terminal differentiation of keratinocytes (Brideau et al. 2007). Tumour lymphangiogenesis was inhibited by endostatin at an early stage in skin tumour development. Inhibition of lymphangiogenesis was

D. Ribatti, E. Crivellato, *Mast Cells and Tumours*,
DOI 10.1007/978-94-007-1469-4_8, © Springer Science+Business Media B.V. 2011

accompanied by significant reduction of lymphatic vessels and lymph node metastasis. Remarkably, accumulation of tumour-infiltrating VEGF-C-positive MCs was markedly decreased in the tumours of the J4 mice. In addition, endostatin inhibited the adhesion and migration of murine MC/9 MCs on fibronectin *in vitro*. These data suggest that endostatin can inhibit tumour lymphangiogenesis by decreasing the VEGF-C levels in the tumours via inhibition of MC migration and adhesion.

# Chapter 9
# Drugs Affecting Mast Cells

Preliminary studies using anti-KIT antibodies (Huang et al. 2008), anti-TNF-α antibodies (Gounaris et al. 2007) or the MC stabilizer disodium cromoglycate (cromolyn) (Soucek et al. 2007) in mouse models demonstrated promising results even if administered after the initiation of tumour development. Treatment of mice bearing mammary adenocarinoma and pancreatic cancer with cromolyn led to clotting of blood vessels, hypoxia and tumour cell apoptosis (Samoszuk and Corwin 2003a, b; Soucek et al. 2007). Unfortunately, cromolyn is a weak inhibitor of human MC secretion and is poorly adsorbed, so it is unlikely to be effective in treating cancer in humans.

At least in humans, MCs seem to depend on the activation of KIT for survival. The first tyrosine kinase inhibitor introduced into the clinic, STI571 (Imatinib mesylate, Gleevec) has inhibitory activity against the signalling cascade activated by KIT receptor (CD117) (Heinrich et al. 2000). This inhibitory activity is the basis of the use of this drug against gastrointestinal stromal tumours (GIST), most of which harbour a KIT mutation (Kitamura and Hirotab 2004). STI571 has been tested in a variety of human tumours, including breast carcinoma, endocrine tumours, ovarian tumours and small cell lung cancer (Cristofanilli et al. 2008; Schilder et al. 2008). None of the clinical trial was successful. The possible causes for these failures are either insufficient MC depletion by STI571 or a non-essential role for MCs in these tumours (Nechushtan 2010). STI571 is also in use for some varieties of mastocytosis, although some KIT activating mutations involved in mastocytosis are resistant to its inhibitory activity (Akin and Metcalfe 2004) and cromolyn has been administered with very limited side effects for various allergic inflammatory conditions (Edwards 1995; Norris 1996; Pardanani et al. 2003).

Ranitidine, a H2R antagonist, used as adjuvant therapy, prolonged survival of colorectal cancer patients (Nielsen et al. 2002). Another H2R antagonist, famotidine, given preoperatively for 14 days, enhanced tumour infiltrating lymphocytes and increased metastatic lymph node reactive changes in breast cancer in humans (Parshad et al. 2002).

Molica et al. (2007) demonstrated that the reduction of the extent of bone marrow angiogenesis observed after sequential therapy with low doses of subcutaneous

D. Ribatti, E. Crivellato, *Mast Cells and Tumours*,
DOI 10.1007/978-94-007-1469-4_9, © Springer Science+Business Media B.V. 2011

alentzumab after a clinical response to fludarabile induction therapy was associated to a reduction in the number of MCs.

Dietary supplementation of silymarin, a naturally derived polyphenolic antioxidant, has been demonstrated to exert a beneficial role in N-nitrosodiethylamine (NDEA)-induced liver cancer in Wistar albino male rats (Ramakrishnan et al. 2009). NDEA administered rats showed increased MC density, as revealed by toluidine staining, along with upregulated expressions of MMP-2 and MMP-9. Silymarin treatment inhibited the increase in MC density and downregulated the expression of MMP-2 and MMP-9 as revealed by Western blotting and immunocytochemistry. Thus, silymarin may exert beneficial effects on liver carcinogenesis by attenuating the recruitment of MCs and thereby decreasing the expressions of MMP-2 and MMP-9 (Ramakrishnan et al. 2009).

The existence of apparent specific MC tryptase inhibitors, such as one used by medicinal leeches, is taken as indirect evidence that tryptases target peptides in pathogens. The first generation of high-potency inhibitors of human tryptases were aromatic bis-amidine, which were used to reduce antigen-induced early and late airway responses in a sheep model of allergic asthma (Caughey et al. 1993; Clark et al. 1995). Later-generation inhibitors of MC tryptases showed similar results in sheep, guinea pig, and mouse models of asthma (Wright et al. 1999; Oh et al. 2002). The tryptase inhibitor APC366 was able to reduce antigen-induced late bronchoconstrictor response in subjects with atopic asthma (Krishna et al. 2001). Another tryptase inhibitor, APC 2059, appeared efficacious in patients with ulcerative colitis (Tremaine et al. 2002). A specific chymase inhibitor, NK3201, has been shown to suppress bleomycin-induced pulmonary fibrosis in hamsters (Sakaguchi et al. 2004). Given the angiogenic potential of both MC tryptase and chymase, drugs that inhibit either proteases might represent a promising tool in the treatment of some tumour types.

# Chapter 10
# Concluding Remarks and Perspectives

As depicted in this survey, MCs represent a class of phylogenetically old cells which are implicated in a broad spectrum of tissue reactions, ranging from inflammation to healing responses. The roles of these cells in normal physiology and in the pathogenesis of several human diseases in addition to allergic reactions have been the subject of increasing interest. As a matter of fact, MCs are now regarded as effective elements in many physiopathological conditions and actively contribute to functional networks leading to specific disease conditions. For many decades their involvement in tumour biology has largely been recognized. Presently, a great bulk of experimental and clinical investigations allows us to think over the complex interaction between MCs and tumours. Most studies reported a positive correlation between the number of MCs and the clinical prognosis in various human tumour types. However, a negative correlation has also been recognized in some instances. These discrepancies have also been found in experimental studies and could be due to differences in the animal model used and the stage of carcinogenesis investigated.

Experimental investigations have partly dissected and clarified the role played by MCs in tumour development and progression (Fig. 10.1). MC precursors are recruited in the tumour microenvironment by chemotactic molecules, such as SCF, liberated by tumour cells. In the tumour context, MCs become activated and operate at different levels, playing multifaceted functions within the signalling network that is established in such unique biological system. First of all, MCs express direct capacity to interfere with tumour cell growth and survival. MCs, indeed, store in their granules or synthesize *de novo* a series of compounds which may exert divergent effects on tumour cells. Histamine, NGF, FGF-2, IL-8 and PDGF, for instance, negatively affect the host by favouring tumour cell growth. By contrast, other MC mediators like IL-1, IL-4, IL-6, TNF-α, TGF-β, $LTB_4$ and chymase exert beneficial effects by stimulating tumour cell inhibition, apoptosis and disruption. Secondly, MCs release a panel of cytokines and chemokines, such as IL-8, MCP-3, MCP-4 and TNF-α, which attract lymphocytes, macrophages and other inflammatory cells in the tumour scenario. Thus, they are key elements in mounting the inflammatory reaction, which is thought to exert important effects on either tumour inhibition or expansion. In addition, the intervention of SCF-activated MCs in the tumour microenvironment increases the amount of IL-17-producing cells in the tumour

D. Ribatti, E. Crivellato, *Mast Cells and Tumours*,

DOI 10.1007/978-94-007-1469-4_10, 

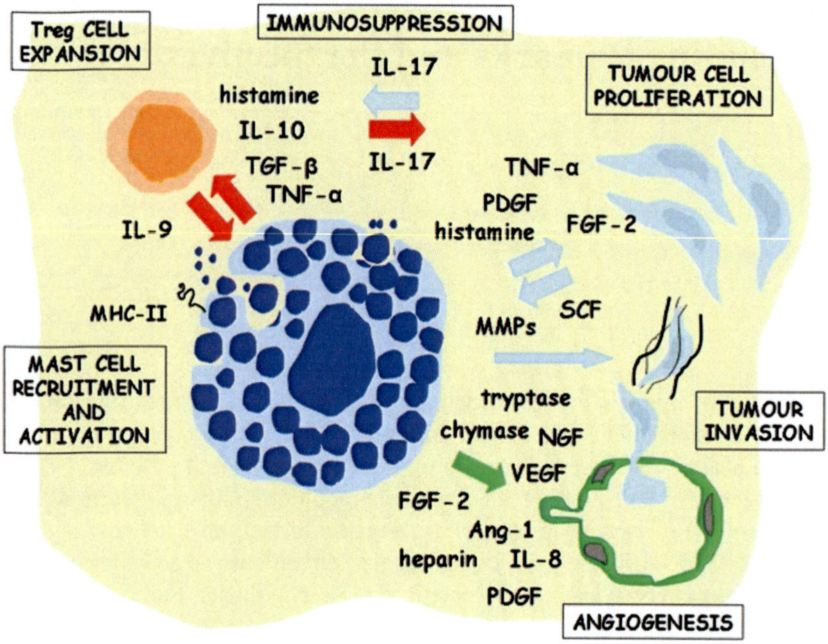


**Fig. 10.1** During tumour development, circulating mast cell precursors migrate into the neoplastic mass and form one of the major stromal cell populations. Recruitment and activation of mast cells in the tumour infiltrate is mainly mediated by tumour-derived SCF. TNF-α, histamine, FGF-2, PDGF and other proinflammatory factors released by mast cells may favour or inhibit the proliferation of tumour cells The intervention of SCF-activated mast cells in the tumour microenvironment increases the amount of IL-17-producing cells in the tumour mass. IL-17, which is a potential candidate for regulating the tumour inflammatory reaction through the production of IL-9, would in turn attract more mast cells at the site of inflammation. Mast cells play a critical role in the suppression of immune reactions by producing inhibitory cytokines, such as IL-10, but also by promoting the immune tolerance mediated by Treg cells. In addition to acting on tumour cells and on the inflammatory infiltrate, mast cells in the tumour setting secrete matrix metalloproteases (MMPs), which favour tumour invasion, and release a number of angiogenic products, which potentiate new blood vessel formation

mass. IL-17, which is a potential candidate for regulating the tumour inflammatory reaction through the production of IL-9, would in turn attract more MCs at the site of inflammation. Besides their direct effect on tumour cells and their involvement in promoting peritumoural inflammatory infiltrates, MCs may play a critical role in the suppression of immune reactions in the tumour setting. This effect is accomplished not only by producing inhibitory cytokines, such as IL-10, but also by promoting the immune tolerance mediated by Treg cells. Next, MCs secrete matrix metalloproteases, such as MMP-2 and MMP-9, which promote extracellular matrix disruption and favour tumour invasion. By contrast, MCs release chondroitin sulfate that embed tumour cells and inhibit metastasis. Finally, MCs are a rich source of potent proangiogenic compounds, such as VEGF, FGF-2 and tryptase, which stimulate blood vessel formation at the periphery of the tumour mass, providing oxygen and nutrient supply as well as escaping routes to tumour cells.

The majority of clinical studies lend support to the concept that a close relationship does exist between the accumulation of MCs in the tumour mass and the poor prognosis of tumour outcome. These data are partly supported by figures from experimental carcinogenesis. Studies in this context suggest that MC intervention is essential for tumour development and progression in distinct types of neoplasia. The introduction of MC-deficient mouse technology in such experimental scenario has demonstrated that some tumours necessitate of MC involvement for their growth. Thus, MCs may operate a key role in both angiogenesis and remodelling of the tumour microenvironment, promoting both tumour initiation and growth. In addition, as tumour growth progresses, MCs may recruit immune cells or alternatively suppress anti-tumoral responses. Uncertainty, however, remains as to the general significance of such findings.

In the light of the present knowledge, MCs might be regarded in a future perspective as a new target for the adjuvant treatment of tumours through the selective inhibition of angiogenesis, tissue remodelling and tumour-promoting molecules, permitting the secretion of cytotoxic cytokines and preventing MC-mediated immune suppression. Moreover, some of the new targeted anti-cancer therapies have pronounced effects on MCs; in fact, it may be that some of their anti-tumour effect is closely related to their effect on MCs.

# References


Aaltomaa S, Lipponen P, Papinaho S et al (1993) Mast cells in breast cancer. Anticancer Res 13:785–788

Abadie A, Besancon F, Wietzerbin J (2004) Type I interferon and TNF alpha cooperate with type II interferon for TRIAL induction and triggering of apoptosis in SK-N-MC EWING tumor cells. Oncogene 23:4911–4920

Abdel-Majid RM, Marshall JS (2004) Prostaglandin E2 induces degranulation-independent production of vascular endothelial growth factor by human mast cells. J Immunol 172:1227–1236

Abramsson A, Lindblom P, Betshowz C (2003) Endothelial and nonendothelial sources of PDGF-B regulate pericyte recruitment and influence vascular pattern formation in tumors. J Clin Invest 112:1142–1151

Acikalin MF, Oner U, Topcu I et al (2005) Tumour angiogenesis and mast cell density in the prognostic assessment of colorectal carcinomas. Dig Liver Dus 37:162–169

Ahmad SA, Liu W, Jung YD et al (2001) The effects of angiopoietin-1 and -2 on tumor growth and angiogenesis in human colon cancer. Cancer Res 61:1255–1259

Akin C, Metcalfe DD (2004) The biology of Kit in disease and the application of pharmacogenetics. J Allergy Clin Immunol 114:13–19

Alberts B, Johnson A, Lewis J et al (2008) Molecular biology of the cell. Garland Science, New York, p 1206

Alcaide P, Jones TG, Lord GM et al (2007) Dendritic cell expression of the transcription factor T-bet regulates mast cell progenitor homing to mucosal tissue. J Exp Med 204:431–439

Alessandri G, Raju KS, Gullino PM (1984) Characterization of a chemoattractant for endothelium induced by angiogenic effectors. Cancer Res 44:1579–1584

Alì G, Boldrini L, Lucchi M et al (2009) Tryptase mast cells in malignant pleural mesothelioma as an independent favorable prognostic factor. J Thorac Oncol 4:348–354

Alitalo K, Tammela T, Petrova TV (2005) Lymphangiogenesis in development and human disease. Nature 438:946–953

Aloe L, Levi-Montalcini R (1977) Mast cells increase in tissues of neonatal rats injected with the nerve growth factor. Brain Res 133:358–366

Andersson CK, Mori M, Bjermer L et al (2009) Novel site-specific mast cell subpopulations in the human lung. Thorax 64:297–305

Anghelina M, Krishnan P, Moldovan L et al (2004) Monocytes and macrophages form branched cell columns in matrigel: implications for a role in neovascularization. Stem Cell Dev 13:665–676

Angiolillo AL, Sgadari C, Taub DD et al (1995) Human interferon-inducible protein 10 is a potent inhibitor of angiogenesis in vivo. J Exp Med 182:155–162

Arinobu Y, Iwasaki H, Gurish MF et al (2005) Developmental checkpoints of the basophil/mast cell lineages in adult murine hematopoiesis. Proc Natl Acad Sci U S A 102:18105–18110

Arinobu Y, Iwasaki H, Akashi K (2009) Origin of basophils and mast cells. Allergol Int 58:21–28



D. Ribatti, E. Crivellato, *Mast Cells and Tumours*,
DOI 10.1007/978-94-007-1469-4, 

Artuc M, Hermes B, Steckelings MU et al (1999) Mast cells and their mediators in wound-healing—active participants or innocent bystanders? Exp Dermatol 8:1–16
Artuc M, Steckelings M, Henz BM (2002) Mast cell-fibroblast interactions: human mast cells as source and inducer of fibroblast and epithelial growth factors. J Invest Dermatol 118:391–395
Asahara T, Takahashi T, Masuda H et al (1999) VEGF contributes to postnatal neovascularization by mobilizing bone marrow-derived endothelial progenitor cells. EMBO J 18:3964–3972
Ashman LK (1999) The biology of stem-cell factor and its receptor c-kit. Int J Biochem Cell Biol 31:1037–1051
Askenase PW (1977) Immune inflammatory responses to parasites: the role of basophils, mast cells and vasoactive amines. Am J Trop Med Hyg 26:96–103
Auerbach R, Alby L, Morrissey LW et al (1985) Expression of organ-specific antigens on capillary endothelial cells. Microvasc Res 29:401–411
Auerbach R, Lu WC, Pardon E et al (1987) Specificity of adhesion between murine tumor cells and capillary endothelium: an in vitro correlate of preferential metastasis in vivo. Cancer Res 47:1492–1496
Azizkhan RG, Azikkhan JC, Zetter BR et al (1980) Mast cell heparin stimulated migration of capillary endothelial cells in vitro. J Exp Med 152:931–944
Babina M, Schülke Y, Kirchhof L et al (2005) The transcription factor profile of human mast cells in comparison with monocytes and granulocytes. Cell Mol Life Sci 62:214–226
Bachelet I, Levi-Schaffer F (2007) Mast cells as effector cells: a costimulating question. Trends Immunol 28:360–365
Bali T, Furth J (1949) A transplantable splenic tumour rich in mast cells: observations on mast cells in varied neoplasms. Am J Pathol 25:605–625
Balkwill F, Mantovani A (2001) Inflammation and cancer: back to Virchow? Lancet 357:539–545
Baram D, Vaday G, Salamon P et al (1991) Human mast cells release metalloproteinase-9 on contact with activated T cells: juxtacrine regulation by TNF-alpha. J Immunol 167:4009–4016
Bashkin P, Razin E, Eldor A et al (1990) Degranulating mast cells secrete an endoglycosidase which degrades heparin sulfate in subendothelial extracellular matrix. Blood 75:2204–2212
Bäumer E (1896) Beiträge zur Histologie der Urticaria simplex und pigmentosa mit besonderer Berüksichtigung der Bedeutung der Mastzellen für die Pathogenese der Urticaria pigmentosa. Arch Derm Syph (Berl) 34:323–348
Beaven MA (2009) Our perception of the mast cell from Paul Ehrlich to now. Eur J Immunol 39:11–25
Behrens M, Taubert M (1952) Der Nachweis von Heparin in den basophilen Leukocyten. Klin Wochenschr 30:76–78
Belloni A, Petrelli L, Guidolin D et al (2006) Identification and localization of adrenomedullin-storing cardiac mast cells. Int J Mol Med 17:709–713
Benditt EP (1956) An enzyme in mast cells with some properties resembling chymotrypsin. Fed Proc 15:507
Benditt EP, Arase M (1959) An enzyme in mast cells with properties like chymotrypsin. J Exp Med 110:451–460
Benditt EP, Wong AL, Arase M et al (1955) 5-hydrpxytryptamine in mast cells. Proc Soc Exp Biol 90:303–304
Benitez-Bribiesca L, Wong A, Utrera D et al (2001) The role of mast cell tryptase in neoangiogenesis of premalignant and malignant lesions of the uterine cervix. J Histochem Cytochem 49:1061–1062
Bergendorff A, Uvnäs B (1973) Storage properties of rat mast cell granules in vitro. Acta Physiol Scand 87:213–222
Bergers G, Brekken R, McMahon G et al (2000) Matrix metalloproteinase-9 triggers the angiogenic switch during carcinogenesis. Nat Cell Biol 2:737–744
Bergers G, Song S, Meyer-Morse N et al (2003) Benefits of targeting both pericytes and endothelial cells in the tumor vasculature with kinase inhibitors. J Clin Invest 111:1287–1295
Betsholtz C (2004) Insight into physiological functions of PGF through genetic studies. Cytokine Growth Factor Rev 15:215–228

Bienenstock J, Befus AD, Denburg J et al (1983) Mast cell heterogeneity. Monogr Allergy 18:124–128
Bingle L, Brown NJ, Lewis CE (2002) The role of tumor associated macrophages in tumor progression; implications for new anticancer therapies. J Pathol 196:254–265
Bischoff J (1995) Approaches to studying cell adhesion molecules in angiogenesis. Trends Cell Biol 5:69–74
Bischoff SC (2007) Role of mast cells in allergic and nonallergic immune response: comparison of human and murine data. Nat Rev Immunol 7:93–104
Bischoff SC (2009) Physiological and pathophysiological functions of intestinal mast cells. Semin Immunopathol 31:185–205
Bischoff SC, Sellge G, Lorentz A et al (1999) IL-4 enhances proliferation and mediator release in mature human mast cells. Proc Natl Acad Sci U S A 96:8080–8085
Black WC, Welch HG (1993) Advances in diagnostic imaging and overestimations of disease prevalence and the benefits of therapy. N Engl J Med 328:1237–1243
Blair RJ, Meng H, Marchese MJ et al (1997) Tryptase is a novel, potent angiogenic factor. J Clin Invest 99:2691–2700
Blank U, Rivera J (2004) The ins and outs of IgE-dependent mast-cell exocytosis. Trends Immunol 25:266–273
Blank U, Ra C, Miller L et al (1989) Complete structure and expression in transfected cells of high affinity IgE receptor. Nature 337:187–189
Bochner BS, Schleimer RP (2001) Mast cells, basophils, and eosinophils: distinct but overlapping pathways for recruitment. Immunol Rev 179:5–15
Bochner BS, Charlesworth EN, Lichtenstein LM et al (1990) Interleukin-1 is released at sites of human cutaneous allergic reactions. J Allergy Clin Immunol 86:830–839
Bockaert J, Claeysen S, Bécamel C et al (2006) Neuronal 5-HT metabotropic receptors: fine-tuning of their structure, signaling, and roles in synaptic modulation. Cell Tissue Res 326:553–572
Boehme SA, Lio FM, Sikora L et al (2004) Cutting edge: serotonin is a chemotactic factor for eosinophils and functions additively with eotaxin. J Immunol 173:3599–3603
Boesiger J, Tsai M, Maurer M et al (1998) Mast cells can secrete vascular permeability factor/vascular endothelial cell growth factor and exhibit enhanced release after immunoglobulin E-dependent upregulation of Fcε receptor I expression. J Exp Med 188:1135–1145
Bonini SE, Lambiase A, Bonini S et al (1996) Circulating nerve growth factor levels are increased in human with allergic diseases and asthma. Proc Natl Acad Sci U S A 93:10955–10960
Bonney V (1908) The connective tissue in carcinoma and in certain inflammatory states that precede its onset. Lancet 1:1389–1393
Bornstein P (1992) Thrombospondins: structure and regulation of expression. FASEB J 6:3290–3299
Bouck N, Stellmach V, Hsu S (1996) How tumors become angiogenic. Adv Cancer Res 69:135–174
Boyce JA, Mellor EA, Perkins B et al (2002) Human mast cell progenitors use α4 integrin, VCAM-1, and PSGL-1 E-selectin for adhesive interactions with human vascular endothelium under flow conditions. Blood 99:2890–2896
Bradding P, Feather IH, Wilson S et al (1993) Immonolocalization of cytokines in the nasal mucosa of normal and perennial rhinitic subjects. The mast cell as a source of IL-4, IL-5, and IL-6 in human allergic mucosal inflammation. J Immunol 151:3853–3865
Bradding P, Okayama Y, Howzrth PH et al (1995) Heterogeneity of human mast cells based on cytokine content. J Immunol 155:297–307
Brideau G, Makinen MJ, Elamaa H et al (2007) Endostatin overexpression inhibits lymhangiogenesis and lymph node metastasis in mice. Cancer Res 67:11528–11535
Brightling CE, Bradding P, Symon FA et al (2002) Mast-cell infiltration of airway smooth muscle in asthma. N Engl J Med 346:1699–1705
Brocklehurst WE (1960) The release of histamine and the formation of slow reacting substance (SRS-A) during anaphylactic shock. J Physiol 151:416–435

Brody D, Metcalfe DD (1998) Mast cells: a unique and functional diversity. Clin Exp Allergy 28:1167–1170
Buckley MG, Gallagher PJ, Walls AF (1998) Mast cell subpopulations in the synovial tissue of patients with osteoarthritis: selective increase in numbers of tryptase-positive, chymase-negative mast cells. J Pathol 186:67–74
Burd PR, Rogers HW, Gordon JR et al (1989) Interleukin 3-dependent and -independent mast cells stimulated with IgE and antigen express multiple cytokines. J Exp Med 170:245–257
Burtin C, Ponvert C, Fray A (1985) Inverse correlation between tumor incidence and tissue histamine levels in W/Wv, W/+, and +/+ mice. J Natl Cancer Inst 74:671–674
Bytautiene E, Vedernikov YP, Saade GR et al (2003) Effect of histamine on phasic and tonic contractions of isolated uterine tissue from pregnant women. Am J Obstetric Gynecol 188:774–778
Cabanillas-Saez A, Schalper JA, Nicovani SM et al (2002) Characterization of mast cells according to their content of tryptase and chymase in normal and neoplastic human uterine cervix. Int J Gynecol Cancer 12:92–98
Cairns JA, Walls AF (1997) Mast cell tryptase stimulates the synthesis of type I collagen in human lung fibroblasts. J Clin Invest 99:1313–1321
Cantarella G, Lempereur L, Presta M et al (2002) Nerve growth factor-endothelial cell interactions lead to angiogenesis in vitro and in vivo. FASEB J 16:1307–1309
Cantor AB, Iwasaki H, Arinobu Y et al (2008) Antagonism of FOG-1 and GATA factors in fate choice for the mast cell lineage. J Exp Med 205:611–624
Cao J, Papadopoulou N, Kempuraj D et al (2005) Human mast cells express corticotropin-releasing hormone (CRH) receptors and CRH leads to selective secretion of vascular endothelial growth factor. J Immunol 174:7665–7675
Cao L, Curtis CL, Theoharides TC (2006) Corticotropin-releasing hormone induces vascular endothelial growth factor release from human mast cells via the cAMP/protein kinase A/p38 mitogen activate protein kinase pathway. Mol Pharmacol 69:998–1006
Cao R, Björndahl MA, Religa P et al (2004) PDGF-BB induces intratumoral lymphangiogenesis and promotes lymphatic metastasis. Cancer Cell 6:333–345
Caplan RM (1963) The natural course of urticaria pigmentosa. Arch Dermatol 87:146–157
Carlini MJ, Daluzzo MCL, Lastiri JM et al (2010) Mast cell phenotypes and microvessels in non-small cell lung cancer and its prognostic significance. Hum Pathol 41:687–705
Carmeliet P, Dor Y, Herbert JM et al (1998) Role of HIF-1 in hypoxia-mediated apoptosis, cell proliferation and tumor angiogenesis. Nature 394:485–490
Cass R, Riley JF, West GB et al (1954) Heparin and histamine in mast cell tumours from dogs. Nature Lond 174:318
Castellot JJ, Karnovsky MJ, Spiegelman BM (1982) Differentiation-dependent stimulation of neovascularization and endothelial cell chemotaxis by 3T3 adipocytes. Proc Natl Acad Sci U S A 79:5597–5601
Castor W (1960) The microscopic structure of normal human synovial tissue. Arthritis Rheum 3:140–151
Catini C, Legnaioli M (1992) Role of mast cells in health: daily rhythmic variations in their number, exocytotic activity, histamine and serotonin content in the rat thyroid gland. Eur J Histochem 36:501–516
Caughey GH, Raymond WW, Bacci E et al (1993) Bis-(5-amidino-2-benzimidazolyl)methane and related amidines are potent, reversible inhibitors of mast cell tryptases. J Pharmacol Exp Ther 264:676–682
Chae SS, Paik JH, Furneaux H et al (2004) Requirement for sphingosine 1-phosphate receptor-1 in tumor angiogenesis demonstrated by in vivo RNA interference. J Clin Invest 114:1082–1089
Chambers AF, Groom AC, Mac Donald IC (2002) Dissemination and growth of cancer cells in metastatic sites. Nat Rev Cancer 2:563–572
Chan JK, Magistris A, Loizzi V et al (2005) Mast cell density, angiogenesis, blood clotting, and prognosis in women with advanced ovarian cancer. Gynecol Oncol 99:20–25
Chang LK, Garcia-Cardeña G, Farnebo F et al (2004) Dose-dependent response of FGF-2 for lymphangiogenesis. Proc Natl Acad Sci U S A 101:11658–11663

Chang YS, di Tomaso E, Mc Donald DM et al (2000) Mosaic blood vessels in tumors: frequency of cancer cells in contact with flowing blood. Proc Natl Acad Sci U S A 97:14608–14613
Chantrain CF, Shimada H, Jodele S et al (2004) Stromal matrix metalloproteinase-9 regulates the vascular architecture in neuroblastoma by promoting pericyte recruitment. Cancer Res 64:1675–1686
Chen CC, Grimbaldeston MA, Tsai M et al (2005) Identification of mast cell progenitors in adult mice. Proc Natl Acad Sci U S A 102:11408–11413
Chen R, Fairley JA, Zhao ML et al (2002) Macrophages, but not T and B lymphocytes, are critical for subepidermal blister formation in experimental bullous pemphigoid: macrophage-mediated neutrophil infiltration depends on mast cell activation. J Immunol 169:3987–3992
Chen W, Jin W, Hardegen N et al (2003) Conversion of peripheral CD4+CD25- naïve T cells to CD4+CD25+ regulatory T cells by TGF-beta induction of transcription factor Foxp3. J Exp Med 198:1875–1886
Chen W, Beck I, Schober W et al (2010) Human mast cells express androgen receptors but treatment with testosterone exerts no influence on IgE-independent mast cell degranulation elicited by neuromuscular blocking agents. Exp Dermatol 19:302–304
Cherwinski HM, Murphy CA, Joyce BL (2005) The CD200 receptor is a novel and potent regulator of murine and human mast cell function. J Immunol 174:1348–1356
Cho SH, Yao Z, Wang SW et al (2003) Regulation of activin A expression in mast cells and asthma: its effect on the proliferation of human airway smooth muscle cells. J Immunol 170:4045–4052
Christofori G, Naik P, Hanahan D (1995) Vascular endothelial growth factor and its receptors, flt-1 and flk-1 are expressed in normal pancreatic islets and throughout islet tumorigenesis. Mol Endocrinol 9:1760–1770
Christy AL, Brown MA (2007) The multitasking mast cell: positive and negative roles in the progression of autoimmunity. J Immunol 179:2673–2679
Church M, Levi-Schaffer F (1997) The human mast cell. J Allergy Clin Immunol 99:155–160
Ch'ng S, Wallis RA, Yuan L et al (2006) Mast cells and cutaneous malignancies. Modern Pathol 19:149–159
Cinel L, Aban M, Bastruk M et al (2009) The association of mast cell density with myometrial invasion in endometrial carcinoma: a preliminary report. Pathol Res Pract 205:255–258
Claffey KP, Brown LF, Del Aguila LF et al (1996) Expression of vascular permeability factor/vascular endothelial growth factor by melanoma cells increases tumor growth, angiogenesis, and experimental metastasis. Cancer Res 56:172–181
Clark JM, Abraham WM, Fishman CE et al (1995) Tryptase inhibitors block allergen-induced airway and inflammatory responses in allergic sheep. Am J Respir Crit Care Med 152:2076–2083
Clowes GH, Owen AG (1904) Metachromatism of mast-cell granules and mucin. J Med Res 7:407–431
Colombo MP, Trinchieri G (2002) Interleukin-12 in anti-tumor immunity and immunotherapy. Cytokine Growth Factor Rev 13:155–168
Columbo M, Bochner BS, Marone G (1995) Human skin mast cells express functional β1 integrins that mediate adhesion to extracellular matrix proteins. J Immunol 154:6058–6064
Conti P, Pang X, Boucher W et al (1997) Impact of Rantes and MCP-1 chemokines on in vivo basohilic mast cell recruitment in rat skin injection model and their role in modifying the protein and mRNA levels for histidine decarboxylase. Blood 89:4120–4127
Coussens LM, Raymond WW, Bergers G et al (1999) Inflammatory mast cells up-regulate angiogenesis during squamous epithelial carcinogenesis. Genes Dev 13:1382–1397
Coussens LM, Werb Z (1996) Matrix metalloproteinases and the development of cancer. Chem Biol 3:895–904
Crapper RM, Schrader JW (1983) Frequency of mast cell precursors in normal tissues determined by an in vitro assay: antigen induces parallel increases in the frequency of P cell precursors and mast cells. J Immunol 131:923–928
Cristofanilli M, Morandi P, Krishnamurthy S et al (2008) Imatinib mesylate (Gleevec) in advanced breast cancer-expressing C-Kit or PDGFR-beta: clinical activity and biological correlations. Ann Oncol 19:1713–1719

Crivellato E, Ribatti D (2010a) The fundamental contribution of William Bate Hardy to shape the concept of mast cell heterogeneity. Br J Haematol 150:152–157
Crivellato E, Ribatti D (2010b) The mast cell: an evolutionary perspective. Biol Rev Camb Philos Soc 85:347–360
Crivellato E, Damiani D, Mallardi F et al (1991) Suggestive evidence for a microanatomical relationship between mast cells and nerve fibres containing substance P, calcitonin gene related peptide, vasoactive intestinal polypeptide, and somatostatin in the rat mesentery. Acta Anat (Basel) 141:127–132
Crivellato E, Candussio L, Decorti G et al (1999) Adryamicin induces exocytosis in rat and beige mouse peritoneal mast cells: an ultrastructural, morphometric and biochemical study. J Submicrosc Cytol Pathol 31:279–286
Crivellato E, Nico B, Vacca A et al (2002a) Mast cell heterogeneity in B-cell non-Hodgkin's lymphomas: an ultrastructural study. Leuk Lymphoma 43:2201–2205
Crivellato E, Candussio L, Mallardi F et al (2002b) Recombinant human alpha-2a interferon promotes an atypical process of mast cell secretion with ultrastructural features suggestive for piecemeal degranulation. J Anat 201:507–512
Crivellato E, Beltrami CA, Mallardi F et al (2003a) Paul Ehrlich's doctoral thesis: a milestone in the study of mast cells. Br J Haematol 123:19–21
Crivellato E, Finato N, Isola M et al (2003b) Low mast cell density in the human duodenal mucosa from chronic inflammatory duodenal bowel disorders is associated with defective villous architecture. Eur J Clin Invest 33:601–610
Crivellato E, Nico B, Vacca A et al (2003c) Ultrastructural analysis of mast cell recovery after secretion by piecemeal degranulation in B-cell non-Hodgkin's lymphoma. Leuk Lymphoma 44:517–521
Crivellato E, Beltrami CA, Mallardi F et al (2004) The mast cell: an active participant or an innocent bystander? Histol Histopathol 19:259–270
Crivellato E, Nico B, Battistig M et al (2005a) The thymus is a site of mast cell development in chicken embryos. Anat Embryol (Berl) 209:243–249
Crivellato E, Finato N, Ribatti D et al (2005b) Do mast cells affect villous architecture? Facts and conjectures. Histol Histopathol 20:1285–1293
Dabbous M, Walker R, Haney L et al (1986) Mast cells and matrix degradation at sites of tumor invasion in rat mammary adenocarcinoma. Br J Cancer 54:459–465
Dabiri S, Huntsman D, Makretsov N et al (2004) The presence of stromal mast cells identifies a subset of invasive breast cancers with a favorable prognosis. Mod Pathol 17:690–695
Dabrowski R, Maslinski C, Gorski P (1975) The effects of histamine liberators and exogenous histamine on wound healing in rat. Agents Actions 5:311–314
Dameron KM, Volpert OV, Tainsky MA et al (1994) Control of angiogenesis in fibroblasts by p53 regulation of thrombospondin-1. Science 265:1582–1584
Damert A, Machein M, Breier G et al (1997) Up-regulation of vascular endothelial growth factor expression in a rat glioma is conferred by two distinct hypoxia-driven mechanisms. Cancer Res 57:3860–3864
Dave SS, Wright G, Than B (2004) Prediction of survival in follicular lymphoma based on molecular features of tumor-infilitrating immune cells. N Engl J Med 351:2159–2169
Davidoff AM, Ng CY, Brown P et al (2001) Bone marrow-derived cells contribute to tumor neovasculature and, when modified to express an angiogenesis inhibitor, can restrict tumor growth in mice. Clin Cancer Res 7:2870–2879
Day M, Green JP (1962) The uptake of biogenic amines by neoplastic mast cells in culture. J Physiol 164:227–237
Demeure CE, Brahimi K, Hacini F et al (2005) Anopheles moschito bites activate cutaneous mast cells leading to a local inflammatory response and lymph node hyperplasia. J Immunol 174:3932–3940
Denhardt DT, Noda M, O'Regan AW et al (2001) Osteopontin as a means to cope with environmental insults: regulation of inflammation, tissue remodeling, and cell survival. J Clin Invest 107:1055–1061

De Palma M, Venneri MA, Naldini L (2003a) In vivo targeting of tumor endothelial cells by systemic delivery of lentiviral vectors. Hum Gene Ther 14:1193–1206
De Palma M, Venneri MA, Roca C et al (2003b) Targeting exogenous genes to tumor angiogenesis by transplantation of genetically modified hematopoietic stem cells. Nat Med 9:789–795
Detoraki A, Staiano RI, Granata F et al (2009) Vascular endothelial growth factors synthesized by human lung mast cells exert angiogenic effects. J Allergy Clin Immunol 123:1142–1149
Dias S, Boyd R, Balkwill F (1998) IL-12 regulates VEGF and MMPs in a murine breast cancer model. Int J Cancer 78:361–365
Di Girolamo N, Wakefiled D (2000) In vitro and in vivo expression of interstitial collagenase/MMP-1 by human mast cells. Dev Immunol 7:131–142
Dimitriadou V, Rouleau A, Dam Trung Tuong M et al (1994) Functional relationship between mast cells and C-sensitive nerve fibres evidenced by histamine H3-receptor modulation in rat lung and spleen. Clin Sci (Lond) 87:151–163
Di Nardo A, Vitiello A, Gallo RL (2003) Cutting edge: mast cell antimicrobial activity is mediated by expression of cathelicidin antimicrobial peptide. J Immunol 170:2274–2278
Dropp JJ (1979) Mast cells in the human brain. Acta Anat (Basel) 105:505–513
Dvorak AM (1991) Basophil and mast cell degranulation and recovery. In: Harris JR (ed) Blood cell Biochemistry, vol 4. Plenum Press, New York, pp 237–344
Dvorak AM (2005a) Piecemeal degranulation of basophils and mast cells is effected by vesicular transport of stored secretory granule contents. Chem Immunol Allergy 85:135–184
Dvorak AM (2005b) Ultrastructural studies of human basophils and mast cells. J Histochem Cytochem 53:1043–1070
Dvorak AM, Kissel S (1991) Granule changes of human skin mast cells characteristic of piecemeal degranulation and associated with recovery during wound healing in situ. J Leuk Biol 49:197–210
Dvorak AM, Morgan ES (2000) Ultrastructural immonogold cytochemistry with autoimmune human sera and an antibody to uridine implicates human mast cell granules in RNA biology. Histochem J 32:685–696
Dvorak AM, Morgan ES, Weller PF (2003) RNA is closely associated with human mast cell lipid bodies. Histol Histopathol 18:943–968
Dvorak HF (1986) Tumors: wounds that not heal. Similarities between tumor stroma generation and wound healing. New Engl J Med 315:1650–1659
Dvorak HF, Dvorak AM (1975) Basophilic leukocytes: structure, function and role in disease. In: Lichtman MA (ed) Clinics in Haematology, vol 4. Saunders WB Co, London, p 651
Dvorak HF, Dvorak AM, Churchill WH (1973) Immunologic rejection of diethylnitrosamine-induced hepatomas in strain 2 guinea pigs. Participation of basophilic leukocytes and macrophages aggregates. J Exp Med 137:751–775
Dvorak HF, Mihm MC Jr, Dvorak AM et al (1974) Morphology of delayed type hypersensitivity reactions in man. I. Quantitative description of the inflammatory response. Lab Invest 31:111–130
Eberhard A, Kahlert S, Goede V et al (2002) Heterogeneity of angiogenesis and blood vessel maturation in human tumors: implications for antiangiogenic tumor therapies. Cancer Res 60:1388–1393
Echternacher B, Männel DN, Hültner L (1996) Critical protective role of mast cells in a model of acute septic peritonitis. Nature 381:75–77
Edwards AM (1995) Oral sodium cromoglycate: its use in the management of food allergy. Clin Exp Allergy 25:31–33
Edwards ST, Cruz AC, Donnelly S et al (2005) c-Kit immunophenotyping and metalloproteinase expression profiles of mast cells in interstitial lung diseases. J Pathol 206:279–290
Egami K, Murohara T, Shimada T et al (2003) Role of host angiotensin II type 1 receptor in tumor angiogenesis and growth. J Clin Invest 112:67–75
Ehrlich P (1878) Beiträge zur Theorie und Praxis der histologiscen Färbung. Thesis, Leipzig University, Leipzig

Ehrlich P (1879) Beiträge zur Kenntnis der granulierten Bindgewebenzellen und der eosinophilen Leukocyten. Archiv Anat Physiol 3:166–169
Ehrlich P (1891) Farbenanalytische Untersuchungen zur Histologie und Klinik des Blutes. Hirschwald, Berlin
Ehrlich P, Lazarus A (1909) Die Anemie, 1. Normale und Patologische Histologie des Blutes. Holder, Wien
Eisman R, Surrey S, Ramachandran B et al (1990) Structural and functional comparison of the genes for human platelet factor 4 and PF4alt. Blood 76:336–344
Elbadawi A (1997) Interstitial cystitis: a critique of current concepts with a new proposal for pathologic diagnosis and pathogenesis. Urology 49:14–40
Ellis LM, Staley CA, Liu W et al (1998) Downregulation of vascular endothelial growth factor in a human colon carcinoma cell line transfected with an antisense expression vector specific for c-src. J Biol Chem 273:1052–1057
Elpek GO, Gelen T, Aksoy NH (2001) The prognostic relevance of angiogenesis and mast cells in squamous cell carcinoma of the oesophagus. J Clin Pathol 54:940–944
Emanueli C, Salis MB, Pinna A et al (2002) Nerve growth factor promotes angiogenesis and arteriogenesis in ischemic hindlimbs. Circulation 106:2257–2262
Enerbäck L (1966a) Mast cells in rat gastrointestinal mucosa. 1. Effects of fixation. Acta Pathologica et Microbiologica Scandinavica 66:289–302
Enerbäck L (1966b) Mast cells in rat gastrointestinal mucosa. 2. Dye-binding and metachromatic properties. Acta Pathologica et Microbiologica Scandinavica 66:303–312
Enerbäck L (1986) Mast cell heterogeneity: the evolution of the concept of a specific mucosal mast cell. In: Befus AD, Bienenstock J, Denburg JA (eds) Mast cell differentiation and heterogeneity. Raven Press, New York, pp 1–26
Erspamer V, Asero B (1952) Identification of enteramine, specific hormone of enterochromaffin cells, as 5-hydroxytryptamine. Nature 169:800–801
Esposito P, Chandler N, Kandere K et al (2002) Corticotropin-releasing hormone and brain mast cells regulate blood-brain-barrier permeability induced by acute stress. J Pharmacol Exp Ther 303:1061–1066
Esposito I, Menucagli M, Funel N et al (2004) Inflammatory cells contribute to the generation of an angiogenic phenotype in pancreatic ductal adenocarcinoma. J Clin Pathol 57:630–636
Estelle F, Simons R (2008) Anaphylaxis. J Allergy Clin Immunol 121:402–407
Fang KC, Wolters PJ, Steinhoff M et al (1999) Mast cell expression of gelatinase A and B is regulated by kit ligand and TGF-β. J Immunol 162:5528–5535
Fang Y, Larsson L, Mattsson J (2010) Mast cells contribute to the mucosal adjuvant effect of CTA1-DD after IgG-complex formation. J Immunol 185:2935–2941
Fawcett DW (1954) Cytological and pharmacological observations on the release of histamine by mast cells. J Exp Med 100:217–224
Féger F, Varadaradjalou S, Gao Z et al (2002) The role of mast cells in host defense and their subversion by bacterial pathogens. Trends Immunol 23:151–158
Fehlner-Gardiner CC, Cao H, Jackson-Boeters L et al (1999) Characterization of a functional relationship between hepatocyte growth factor and mouse bone marrow-derived mast cells. Differentiation 65:27–42
Feldberg W, Kellaway DH (1938) Liberation of histamine and formation of lyscithin-like substances by cobra venom. J Physiol 94:187–226
Feoktistov I, Ryzhov S, Glodstein AE et al (2003) Mast cell-mediated stimulation of angiogenesis. Cooperative interaction between A2 and A3 adenosine receptors. Circ Res 92:485–492
Fernandez A, Udagawa T, Schwesinger C et al (2001) Angiogenic potential of prostate carcinoma cells overexpressing bcl-2. J Natl Cancer Inst 93:208–213
Fernandez Pujol B, Lucibello FC, Gehling UM et al (2000) Endothelial-like cells derived from human CD14 positive monocytes. Differentiation 65:287–300
Ferrara N (2007) Vascular endothelial growth factor: pathophysiology and clinical implications. In: Ferrara N (ed) Angiogenesis. From basic science to clinical applications. CRC, New York, pp 1–36

Fidler IJ (1986) Rationale and methods for the use of nude mice to study the biology and therapy of human cancer metastasis. Cancer Metastasis Rev 5:29–49
Fidler IJ, Kripke ML (1977) Metastasis results from preexisting variant cells within a malignant tumor. Science 197:893–895
Fidler IJ, Gruys E, Cifone MA et al (1981) Demonstration of multiple phenotypic diversity in a murine melanoma of recent origin. J Natl Cancer Inst 67:947–956
Flynn EA, Schwartz JL, Shklar G (1991) Sequential mast cell infiltration and degranulation during experimental carcinogenesis J Cancer Res Clin Oncol 117:115–122
Folkman J (1971) Tumour angiogenesis: therapeutic implications. N Engl J Med 285:1182–1186
Folkman J, Shing Y (1992) Angiogenesis. J Biol Chem 267:10931–10934
Folkman J, Watson K, Ingber D et al (1989) Induction of angiogenesis during the transition from hyperplasia to neoplasia. Nature 339:58–61
Foreman JC, Jordan CC (1980) Comparison of the activities of some peptides in human skin and as histamine releasing agents. Proc Br Pharmacol Soc (Lond) C 57:55
Forsberg E, Pejler G, Ringvall M et al (1999) Abnormal mast cells in mice deficient in a heparin-synthesizing enzyme. Nature 400:773–776
Frandji P, Tkaczyk C, Oskeritzian C et al (1996) Exogenous and endogenous antigens are differentially presented by mast cells to CD4+ T lymphocytes. Eur J Immunol 26:2517–2528
Frangogiannis NG, Lindsey ML, Michael LH et al (1998) Resident cardiac mast cells degranulate and release preformed TNF-α, initiating the cytokine cascade in experimental canine myocardial ischemia/reperfusion. Circulation 98:699–710
Frank BT, Rossall JC, Caughey GH et al (2001) Mast cell tissue inhibitor of metalloproteinase-1 is cleaved and inactivated extracellularly by alpha-chymase. J Immunol 166:2783–2792
Fraser RA, Simpson JG (1980) Role of mast cells in experimental tumour angiogenesis. In: Nugent J, O'Connor M (eds) Development of the vascular system. Ciba foundation symposium 100, vol 1983. Pitman, London, pp 120–131
Fröberg GK, Lindberg R, Ritter M et al (2009) Expression of serotonin and its 5-HT1A receptor in canine cutaneous mast cell tumours. J Comp Pathol 141:89–97
Fromme E (1906) Demonstration über das Verhalten der Mastzellen beim Karzinom. Zbl Gynäk 30:1146–1147
Frossi B, Gri G, Tripodo C et al (2010) Exploring a regulatory role for mast cells: 'MCregs'? Trends Immunol 31:97–102
Fukushima N, Satoh T, Sano M (2001) Angiogenesis and mast cells in non-Hodgkin's lymphoma: strong correlation in angioimmunoblastic T-cell lymphoma. Leuk Lymphoma 42:709–720
Fukushima H, Ohsawa M, Ikura Y et al (2006) Mast cells in diffuse large B-cell lymphoma; their role in fibrosis. Histopathology 49:498–505
Furano AV, Green JP (1964) The uptake of biogenic amines by mast cells of the rat. J Physiol 170:263–271
Galli SJ, Lantz CS (1999) Allergy. In: Paul WE (ed) Fundamental immunology. Lippincot-Raven, Philadelphia, pp 1137–1184
Galli SJ, Kalesnikoff J, Grimbaldeston MA et al (2005a) Mast cells as "tunable" effector and immunoregulatory cells: recent advances. Annu Rev Immunol 23:749–786
Galli SJ, Nakae S, Tsai M (2005b) Mast cells in the development of adaptive immune responses. Nat Immunol 6:135–142
Galli SJ, Grimbaldeston M, Tsai M (2008a) Immunomodulatory mast cells: negative, as well as positive, regulators of immunity. Nat Rev Immunol 8:445–454
Galli SJ, Tsai M, Piliponski AM (2008b) The development of allergic inflammation. Nature 454:445–454
Gaudenzio N, Espagnolle N, Mars LT et al (2009) Cell-cell cooperation at the T helper cell/mast cell immunological synapse. Blood 114:4979–4988
Gately S, Twardowski P, Tack MS et al (1997) The mechanism of cancer-mediated conversion of plasminogen to the angiogenesis inhibitor angiostatin. Proc Natl Acad Sci U S A 94:10868–10872

Genovese A, Rossi FW, Spadaro G et al (2010) Human cardiac mast cells in anaphylaxis. Chem Immunol Allergy 95:98–109

Ghildyal N, McNeil HP, Gurish MF et al (1992) transcriptional regulation of the mucosal mast cell-specific protease gene, MMCP-2, by interleukin 10 and interleukin 3. J Biol Chem 267:8473–8477

Giancotti FG, Ruoslahti E (1999) Integrin signaling. Science 285:1028–1032

Gilfillan AM, Tkaczyk C (2006) Integrated signalling pathways for mast-cell activation. Nat Rev Immunol 6:218–230

Giraudo E, Inoue M, Hanahan D (2004) An amino-bisphosphonate targets MMP-9-expressing macrophages and angiogenesis to impair cervical carcinogenesis. J Clin Invest 114:623–633

Glenner GG, Cohen LA (1960) Histochemical demonstration of a species-specific trypsin-like enzyme in mast cells. Nature 185:846–847

Glimelius I, Edstrom A, Fischer M et al (2005) Angiogenesis and mast cells in Hodgkin lymphoma. Leukemia 19:2360–2362

Gombert M, Dieu-Nosjean MC, Winterberg F et al (2005) CCL1-CCR8 interactions: an axis mediating the recruitment of T cells and Langerhans-type dendritic cells to sites of atopic skin inflammation. J Immunol 174:5082–5091

Gomes AP, Johann JE, Lovato GG et al (2008) Comparative analysis of the mast cell density in normal oral mucosa, actinic cheilitis and lip squamous cell carcinoma. Braz Dent J 19:186–189

Gommerman JL, Oh DY, Zhou X et al (2000) A role for CD21/CD35 and CD19 in responses to acute septic peritonitis: a potential mechanism for mast cell activation. J Immunol 165:6915–6921

Gomori G (1953) Chloroacyl esters as histochemical substrates. J Histochem Cytochem 1:469–470

Gong J, Yang NS, Croft M et al (2010) The antigen presentation function of bone marrow-derived mast cells is spatiotemporally restricted to a subset expressing high levels of cell surface FcepsilonRI and MHC II. BMC Immunol 11:34

Gonzalez S, Moran M, Kochevar IE (1999) Chronic photodamage in skin of mast cell-deficient mice. Photochem Photobiol 70:248–253

Gooch SJ, Lee AV, Yee D (1998) Interleukin 4 inhibits growth and induces apoptosis in human breast cancer cells. Cancer Res 58:4199–4205

Good DJ, Polverini PJ, Rastinejad F et al (1990) A tumor suppressor-dependent inhibitor of angiogenesis is immunologically and functionally indistinguishable from a fragment of thrombospondin. Proc Natl Acad Sci U S A 87:6624–6628

Gordon JR, Galli SJ (1990) Mast cells as a source of both preformed and immunologically inducible TNF-α/cachetin. Nature 346:274–276

Gordon JR, Galli SJ (1991) Release of both preformed and newly synthesized tumor necrosis factor alpha (TNF-α)/cachectin by mouse mast cells stimulated via the FcεRI. A mechanism for the sustained action of mast cell-derived TNF-α during IgE-dependent biological responses. J Exp Med 174:103–107

Gottfried E, Kreutz M, Haffner S et al (2007) Differentiation of human tumour-associated dendritic cells into endothelial-like cells: an alternative pathway of tumour angiogenesis. Scand J Immunol 65:329–335

Gounaris E, Erdman SE, Restaino C et al (2007) Mast cells are an essential hematopoietic component for polyp development. Proc Natl Acad Sci U S A 104:19977–19982

Graeber TG, Osmanian C, Jacks T et al (1996) Hypoxia-mediated selection of cells with diminished apoptotic potential in solid tumours. Nature 379:88–91

Graham HT, Lowry OH, Wheelwright F et al (1955) Distribution of histamine among leukocytes and platelets. Blood 10:467–481

Greene HS, Harvey EK (1964) The relationship between the dissemination of tumor cells and the distribution of metastases. Cancer Res 24:799–811

Gregory GD, Raju SS, Winandry S et al (2006) Mast cell IL-4 expression is regulated by Ikaros and influences encephalitogenic TH 1 responses in mice. J Clin Invest 116:1327–1336

Gri G, Piconese S, Frossi B et al (2008) CD4+CD25+ regulatory T cells suppress mast cell degranulation and allergic responses through OX40-OX40L interaction. Immunity 29:771–781

Grimbaldeston MA, Chen CC, Piliponski AM et al (2005) Mast cell-deficient W-sash c-kit mutant KitW-sh/W-sh mice as a model for investigating mast cell biology in vivo. Am J Pathol 167:835–848
Grimbaldeston MA, Metz M, Yu M et al (2006) Effector and potential immunoregulatory roles of mast cells in IgE-associated acquired immune responses. Curr Opin Immunol 18:751–760
Grimbaldeston MA, Nakae S, Kalesnikoff K et al (2007) Mast cell-derived interleukin 10 limits skin pathology in contact dermatitis and chronic irradiation with ultraviolet B. Nat Immunol 38:1095–1104
Gruber BL, Marchese MJ, Suzuki K et al (1989) Synovial procollagenase activation by human mast cell tryptase dependence upom matrix metalloproteinase 3 activation. J Clin Invest 84:1657–1662
Gruber BL, Marchese MJ, Kew RR (1994) Transforming growth factor- β1 mediates mast cell chemotaxis. J Immunol 152:5860–5867
Gruber BL, Marchese MJ, Kaw R (1995) Angiogenic factors stimulate mast cell migration. Blood 86:2488–2493
Gruber BL, Kew RR, Jelaska A et al (1997) Human mast cells activate fibroblasts. J Immunol 158:2310–2317
Grutzkau A, Kruger-Krasagakes S, Baumesteir H et al (1998) Synthesis, storage, and release of vascular endothelial growth factor/vascular permeability factor (VEGF/VPF) by human mast cells: implications for the biological significance of VEGF 206. Mol Biol Cell 9:875–884
Guidolin D, Crivellato E, Nico B et al (2006) An image analysis of the spatial distribution of perivascular mast cells in human melanoma. Int J Mol Med 17:981–987
Guidolin D, Nico B, Crivellato E et al (2009) Tumoral mast cells exhibit a common spatial distribution. Cancer Letters 273:80–85
Gulobova M, Vlaykova T (2009) Prognostic significance of mast cell number and microvascular density for the survival of patients with primary colorectal cancer. J Gastroenetrol Hepatol 24:1265–1275
Guo P, Hu B, Gu W et al (2003) Platelet-derived growth factor-B enhances glioma angiogenesis by stimulating vascular endothelial growth factor expression in tumor endothelia and by promoting pericyte recruitment. Am J Pathol 162:1083–1093
Gurish MF, Austen KF (2001) The diverse role of mast cells. J Exp Med 194:1–5
Gurish MF, Boyce JA (2006) Mast cells: ontogeny, homing, and recruitment of a unique innate effector cell. J Allergy Clin Immunol 117:1285–1291
Gurish MF, Ghildyal N, McNeil HP et al (1992) Differential expression of secretory granule proteases in mouse mast cells exposed to interleukin 3 and c-kit ligand. J Exp Med 175:1003–1012
Gurish MF, Tao H, Abonia JP et al (2001) Intestinal mast cell progenitors require CD49dbeta7 (alpha4beta7 integrin) for tissue-specific homing. J Exp Med 194:1243–1252
Gutschalk CM, Herold-Mende CC, Fusenig NE et al (2006) Granulocyte colony-stimulating factor and granulocyte-macrophage colony stimulating factor promote malignant growth of cells from head and neck squamous cells carcinomas in vivo. Cancer Res 66:8026–8036
Guy-Grand D, Dy M, Luffau G et al (1984) Gut mucosal mast cells. Origin, traffic and differentiation. J Exp Med 160:12–28
Halaban R, Kwon BS, Ghost S et al (1988) bFGF as an autocrine growth factor for human melanomas. Oncogene Res 3:177–186
Hanada T, Nakagawa M, Emoto A et al (2000) Prognostic value of tumor-associated macrophage count in human bladder cancer. Int J Urol 7:263–269
Hanahan D (1985) Heritable formation of pancreatic beta-cell tumors in transgenic mice expressing recombinant unsylin/simian virus 40 oncogene. Nature 315:115–122
Hanahan D, Folkman J (1996) Patterns and emerging mechanisms of the angiogenic switch during tumorigenesis. Cell 86:353–364
Hanahan D, Weinberg RA (2000) The hallmarks of cancer. Cell 100:57–70
Hara M, Matsumori A, Ono K et al (1999) Mast cells cause apotosis of cardiomyocytes and proliferation of other intramyocardial cells in vitro. Circulation 100:1443–1449

Hardy WB, Wesbrook FF (1895) The wandering cells of the alimentary canal. J Physiol 18:490–524

Harris HF (1900) Histology and microchemic reactions of some cells to anilin dyes—identity of the plasma-cell and osteoblast-fibrous tissue a secretion of the plasma-cells- mast cells elaborates mucin of connective tissues. Philad Med 5:757–764

Hart IR, Fidler IJ (1980) Role of organ selectivity in the determination of metastatic patterns of B16 melanoma. Cancer Res 40:2281–2287

Hart PH, Grimbaldeston MA, Finlay-Jones JJ (2000) Mast cells in UV-B-induced immunosuppression. J Photochem Photobiol B 55:81–87

Hart PH, Grimbaldeston M, Finlay-Jones J (2001) Immunosuppression and skin cancer: role of histamine and mast cells. Clin Exp Pharmacol Physiol 28:1–8

Hart PH, Townley S, Grimbaldeston M et al (2002) Mast cells, neuropeptides, histamine and prostaglandins in UV-induced systemic immunosuppression. Methods 28:79–89

Harvima IT, Nilsson G, Naukkarinen A (2010) Role of mast cells and sensory nerves in skin inflammation. G Ital Dermatol Venereol 145:195–204

He S, Walls AF (1998) Human mast cell chymase induces the accumulation of neutrophils, eosinophils and other inflammatory cells in vivo. Br J Pharmacol 125:1491–1500

Heinrich MC, Griffith DJ, Druker BJ et al (2000) Inhibition of c-kit receptor tyrosine kinase activity by STI 571 a selective tyrosine kinase inhibitor. Blood 96:925–932

Heissig B, Rafii S, Akiyama H et al (2005) Low-dose irradiation promotes tissue revascularization through VEGF release from mast cells and MMP-9-mediated progenitor cell mobilization. J Exp Med 202:739–750

Hershko AY, Rivera J (2010) Mast cell and T cell communication; amplification and control of adaptive immunity. Immunol Lett 128:98–104

Higuchi K (1930) Die Gewebsmastzellen in der Mamma. Folia haematol 41:401–414

Ho LH, Ohno T, Oboki K et al (2007) IL-33 induces IL-13 production by mouse mast cells independently of IgE-FcepsilonRI signals. J Leukoc Biol 82:1481–1490

Hochegger K, Siebenhaar F, Vielhauer V et al (2005) Role of mast cells in experimental anti-glomerular basement membrane glomerulonephritis. Eur J Immunol 35:3074–3082

Hofmann AM, Abraham SN (2009) New roles form mast cells in modulating allergic reactions and immunity against pathogens. Curr Opin Immunol 21:679–686

Hofstra CL, Desai PJ, Thurmond RL et al (2003) WP: histamine H4 receptor mediates chemotaxis and calcium mobilization of mast cells. J Pharmacol Exp Ther 305:1212–1221

Holash J, Maisonpierre PC, Compton D et al (1999a) Vessel cooption, regression and growth in tumors mediated by angiopoietins and VEGF. Science 284:1994–1998

Holash J, Wiegand SJ, Yancopoulos GF (1999b) New model of tumor angiogenesis: dynamic balance between vessel regression and growth mediated by angiopoietins and VEGF. Oncogene 18:5356–5362

Holgate ST (2002) Airway inflammation and remodeling in asthma: current concepts. Mol Biotechnol 22:179–189

Holmgren H (1946) Beitrag zur Frage der Genese der Ehrlichschen Mastzellen. Acta Anat 2:40–56

Holmgren H, Wilander O (1937) Beitrag zur Kenntniss der Chemie und Funktion der Ehrlichschen Mastzellen. Z Mikrosk Anat Forsch 42:242–278

Holmgren H, Wohlfart G (1947) Mast cells in experimental rat sarcomas. Cancer Res 7:686–691

Huang B, Lei Z, Zhang GM et al (2008) SCF-mediated mast cell infiltration and activation exacerbate the inflammation and immunosuppression in tumor microenvironment. Blood 112:1269–1279

Huang C, Sali A, Stevens RL (1998) Regulation and function of mast cell proteases in inflammation. J Clin Immunol 18:169–183

Huang C, De Sanctis GT, O'Brien PJ et al (2001) Evaluation of the substrate specificity of human mast cell tryptase βI and demonstration of its importance in bacterial infections of the lung. J Biol Chem 276:26276–26284

Huang JF, Thurmond RL (2008) The new biology of histamine receptors. Curr Allergy Asthma Rep 8:21–27

Hujanen ES, Terranova VP (1985) Migration of tumor cells to organ-derived chemoattractants. Cancer Res 45:3517–3521
Humphries DE, Wong GW, Friend DS et al (1999) Heparin is essential for the storage of specific granule proteases in mast cells. Nature 400:769–772
Humrich JY, Humrich JH, Averbeck M et al (2006) Mature monocyte-derived dendritic cells respond more strongly to CCL19 than to CXCL12: consequences for directional migration. Immunology 117:238–247
Iamaroon A, Pongsiriwet S, Jittidecharaks S et al (2003) Increase of mast cells and tumor angiogenesis in oral squamous cell carcinoma. J Oral Pathol Med 32:195–199
Iba O, Matsubara H, Nozawa Y et al (2002) Angiogenesis by implantation of peripheral blood mononuclear cells and platelets into ischemic limbs. Circulation 106:2019–2025
Ibaraki T, Muramatsu M, Takai S et al (2005) The relationship of tryptase- and chymase-positive mast cells to angiogenesis in stage I non-small cell lung cancer. Eur J Cardiothorac Surg 28:617–621
Iikura M, Suto H, Kajiwara N et al (2007) IL-33 can promote survival, adhesion and cytokine production in human mast cells. Lab Invest 87:971–978
Imada D, Shijubo N, Kojima H et al (2000) Mast cells correlate with angiogenesis and poor outcome in stage I lung adenocarcinoma. Eur Resp J 15:1087–1093
Inoue M, Hager JH, Ferrara N et al (2002) VEGF-A has a critical, nonreduntant role in angiogenic switching and pancreatic beta cell carcinogenesis. Cancer Cell 1:193–202
Irani AM, Schwartz LB (1994) Human mast cell heterogeneity. Allergy Proc 15:303–308
Irani AM, Schechter NM, Craig SS et al (1986) Two types of human mast cells that have distinct neutral protease composition. Proc Nat Acad Sci U S A 83:4464–4468
Irani AM, Bradford TR, Kepley CL et al (1989) Detection of MCT and MCTC types of human mast cells by immunohistochemistry using new monoclonal anti-tryptase and anti-chymase antibodies. J Histochem Cytochem 37:1509–1515
Irani AM, Goldstein SM, Wintroub BU et al (1991) Human mast cell carboxypeptidase: selective location to MCTC cells. J Immunol 147:247–253
Ishizaka K, Ishizaka T (1967) Identification of γE-antibodies as a carrier of reaginic reactivity. J Immunol 99:1187–1198
Ishizaka T, Ishizaka K, Orange RP et al (1970) The capacity of human immunoglobulin E to mediate the release of histamine and slow reacting substance of anaphylaxis (SRS-A) from monkey lung. J Immunol 104:335–343
Ito N, Sugawara K, Bodó E et al (2010) Corticotropin-releasing hormone stimulates the in situ generation of mast cells from precursors in the human hair follicle mesenchyme. J Invest Dermatol 130:995–1004
Iwasaki H, Mizuno S, Arinobu Y et al (2006) The order of expression of transcription factors directs hierarchical specification of hematopoietic lineages. Genes Dev 20:3010–3021
Jain RK, Munn LL (2000) Leaky vessels? Call Ang1! Nat Med 6:131–132
Jakobson AM, Hahnenberger R (1991) Antiangiogenic effect of heparin and other sulphated glycosaminoglycans in the chick embryo chorioallantoic membrane. Pharmacol Toxicol 69:122–126
Jakobson A, Sorbo J, Norrby K (1990) Protamine and mast cell-mediated angiogenesis in the rat. Int J Exp Pathol 71:209–217
Javed NH, Wang YZ, Cooke HJ (1992) Neuroimmune interactions: role for cholinergic neurons in intestinal anaphylaxis. Am J Physiol 263:847–852
Jawdat DM, Rowden G, Marshall JS (2006) Mast cells have a pivotal role in TNF-independent lymph node hypertrophy and the mobilization of Langerhans cells in response to bacterial peptidoglycan. J Immunol 177:1755–1762
Jenkins DC, Charles IG, Thomsen LL et al (1995) Role of nitric oxide in tumor growth. Proc Natl Acad Sci U S A 92:4392–4396
Johansson A, Rudolf S, Hammarsten P et al (2010) Mast cells are novel independent prognostic markers in prostate cancer and represent a target for therapy. Am J Pathol 177:1031–1041

Johnson AR, Hugli TE, Müller-Eberhard HJ (1975) Release of histamine from rat mast cells by the complement peptides C3a and C5a. Immunology 28:1067

Johnson D, Krenger W (1992) Interactions of mast cells with the nervous system—recent advances. Neurochem Res 17:939–951

Jolly MJ (1900) Clasmatocytes et mastzellen. Compte Rendus Société de Biologie (Paris) 52:437–455

Jorpes E, Holmgren H, Wilander O (1937) Über das Vorkommen von Heparin in den Gefässwänden und in den Augen. Ein Beitrag zur Physiologie der Ehrlichschen Mastzellen. Z Mikrosk Anat Forsch 42:279–301

Joyce JA, Laakkonen P, Bernasconi M et al (2003) Stage-specific markers revealed by phage display in a mouse model of pancreatic islet tumorigenesis. Cancer Cell 4:393–403

Joyce JA, Baruch A, Chelade K et al (2004) Cathepsin cysteine proteases are effectors on invasive growth and angiogenesis during multistage tumorigenesis. Cancer Cell 5:443–453

Ju MJ, Qiu SJ, Gao Q et al (2009) Combination of peritumoral mast cells and T-regulatory cells predicts prognosis of hepatocellular carcinoma. Cancer Sci 100:1267–1274

Jutel M, Akdis M, Akdis CA (2009) Histamine, histamine receptors and their role in immune pathology. Clin Exp Allergy 39:1786–1800

Kageyama-Yahara N, Suehiro Y, Yamamoto T et al (2008) IgE-induced degranulation of mucosal mast cells is negatively regulated via nicotinic acetylcholine receptors. Biochem Biophys Res Commun 377:321–325

Kajiwara N, Sasaki T, Bradding P et al (2010) Activation of human mast cells through the platelet-activating factor receptor. J Allergy Clin Immunol 125:1137–1145

Kakurai M, Monteforte R, Suto H et al (2006) Mast cell-derived tumor necrosis factor can promote nerve fiber elongation in the skin during contact hypersensitivity in mice. Am J Pathol 169:1713–1721

Kambayashi T, Allenspach EJ, Chang JT et al (2009) Inducible MHC class II expression by mast cells supports effector and regulatory T cell activation. J Immunol 182:4686–4695

Kanamaru Y, Scandiuzzi L, Essig M et al (2006) Mast cell-mediated remodeling and fibrinolytic activity protect against fatal glomerulonephritis. J Immunol 176:5607–5615

Kanbe N, Kurosawa M, Miyachi Y et al (2000) Nerve growth factor prevents apoptosis of cord blood-derived human cultured mast cells synergistically with stem cell factor. Clin Exp Allergy 30:1113–1120

Kandel J, Bossy-Wetzel E, Radvany F et al (1991) Neovascularization is associated with a switch to the export of bFGF in the multistep development of fibrosarcoma. Cell 66:1095–1104

Kanerva K, Lappalainen J, Mäkitie LT et al (2009) Expression of antizyme inhibitor 2 in mast cells and role of polyamines as selective regulators of serotonin secretion. PLoS One 4:e6858

Kankkunen JP, Harvima IT, Naukkarinen A (1997) Qunatitative analysis of tryptase and chymase containing mast cells in benign and malignant breast lesions. Int J Cancer 72:385–388

Kanthack AA, Hardy WB (1894) The morphology and distribution of the wandering cells of mammalia. J Physiol 17:81–119

Karra L, Berent-Maoz B, Ben-Zimra M et al (2009) Are we ready to downregulate mast cells? Curr Opin Immunol 21:708–714

Kay LJ, Peachell PT (2005) Mast cell beta2-adrenoreceptors. Chem Immunol Allergy 87:145–153

Kenny PA, Mc Donald PJ, Finlay-Jones JJ (1993) The effect of cytokines on bactericidal activity of murine neutrophils. FEMS Immunol Med Microbiol 7:271–279

Kerbel R, Folkman J (2002) Clinical translation of angiogenesis inhibitors. Nat Rev Cancer 2:727–739

Kessler DA, Langer RS, Pless NA et al (1976) Mast cells and tumor angiogenesis. Int J Cancer 18:703–709

Khalil Z, Townley SL, Grimbaldeston MA et al (2001) cis-Urocanic acid stimulates neuropeptide release from peripheral sensory nerves. J Invest Dermatol 117:886–891

Khalil M, Ronda J, Weintraub M et al (2007) Brain mast cell relationship to neurovasculature during development. Brain Res 1171:18–29

Khan MM, Strober S, Melmon KL (1986) Regulatory effects of mast cells on lymphoid cells: the role of histamine type 1 receptors in the interaction between mast cells, helper T cells and natural suppressor cells. Cell Immunol 103:41–53

Kim GY, Lee JW, Ryu HC et al (2010) Proinflammatory cytokine IL-1β stimulates IL-8 synthesis in mast cells via a leukotriene B4 receptor 2 linked pathway, contributing to angiogenesis. J Immunol 184:3946–3954

Kindt F, Wiegand S, Niemeier V et al (2008) Reduced expression of nicotinic alpha subunits 3, 7, 9 and 10 in lesional and nonlesional atopic dermatitis skin but enhanced expression of alpha subunits 3 and 5 in mast cells. Br J Dermatol 159:847–857

King CL, Xianli J, Malhotra I et al (1997) Mice with a targeted deletion of the IgE gene have increased worm burdens and reduced granulomatous inflammation following primary infection with Schistosoma mansoni. J Immunol 158:294–300

Kirsch M, Schackert G, Black PM (2004) Metastasis and angiogenesis. Cancer Treat Res 117:285–304

Kirschenbaum AS, Kessel SW, Goff JP et al (1991) Demonstration of the origin of human mast cells from CD34+ bone marrow progenitor cells. J Immunol 146:1410–1415

Kitamura T, Kometani K, Hashida H et al (2007) SMAD4-deficient intestinal tumors recruit CCR1+ myeloid cells that promote invasion. Nat Genet 39:467–475

Kitamura Y, Hirotab S (2004) Kit as a human oncogenic tyrosine kinase. Cell Mol Life Sci 61:2924–2931

Kitamura Y, Ito A (2005) Mast cell-committed progenitors. Proc Natl Acad Sci U S A 102:11129–11130

Kitamura Y, Go S, Hatanaka K (1978) Decrease of mast cells in W/Wv mice and their increase by bone marrow transplantation. Blood 52:447–452

Klimp AH, Hollema H, Kempinga C et al (2001) Expression of cyclooxygenase-2 and inducible nitric oxide synthase in human ovarian tumors and tumor-associated macrophages. Cancer Res 61:7305–7309

Knight PA, Wright SH, Lawrence CE et al (2000) Delayed expulsion of the nematode Trichinella spiralis in mice lacking the mucosal mast cell-specific granule chymase, mouse mast cell protease-1. J Exp Med 192:1849–1856

Koide N, Nishio A, Sato T et al (2004) Significance of macrophage chemoattractant protein-1 expression and macrophage infiltration in squamous cell carcinoma of the esophagus. Am J Gastroenterol 99:1667–1674

Kopicky-Burd JA, Kagey-Sobotka A, Peters SP et al (1988) Characterization of human synovial mast cells. J Rheumatol 15:1326–1333

Kops SK, Van Loveren H, Rosenstein RW et al (1984) Mast cell activation and vascular alterations in immediate hypersensitivity-like reactions induced by a T-cell derived antigen-binding factor. Lab Invest 50:421–434

Koskivirta I, Rahkonen O, Mäyränpää M et al (2006) Tissue inhibitor of metalloproteinases 4 (TIMP4) is involved in inflammatory processes of human cardiovascular pathology. Histochem Cell Biol 126:335–342

Kreuter Kops S, Theoharides TC, Cronin CT et al (1990) Ultrastructural characteristics of rat peritoneal mast cells undergoing differential release of serotonin without histamine and without degranulation. Cell Tissue Res 262:415–424

Kripke M (1984) Immunological unresponsiveness induced by ultraviolet radiation. Immunol Rev 80:87–102

Krishna MT, Chauhan A, Little L et al (2001) Inhibition of mast cell tryptase by inhaled APC 366 attenuates allergen-induced late-phase airway obstruction in asthma. J Allergy Clin Immunol 107:1039–1045

Krop M, Ozünal ZG, Chai W et al (2010) Mast cell degranulation mediates bronchoconstriction via serotonin and not via renin release. Eur J Pharmacol 640:185–189

Kubo H, Cao R, Brakenhielm E et al (2002) Blockade of vascular endothelial growth factor receptor-3 signaling inhibits fibroblast growth factor-2-induced lymphangiogenesis in mouse cornea. Proc Natl Acad Sci U S A 99:8868–8873

Kulka M, Alexopoulou L, Flavell RA et al (2004) Activation of mast cells by double-stranded RNA: evidence for activation through Toll-like receptor 3. J Allergy Clin Immunol 114: 174–182

Kushnir-Sukhov NM, Brown JM, Wu Y et al (2007) Human mast cells are capable of serotonin synthesis and release. J Allergy Clin Immunol 119:498–499

Lago Costa N, Ferreira Oton-Lelte A, Pelxodo Chelm-Júnlor A et al (2009) Density and migration of mast cells in lip squamous cell carcinoma and actinic cheilitis. Histol Histopathol 24: 457–465

Lantz CS, Boesinger J, Song CH et al (1998) Role for interleukin-3 in mast-cell and basophile development and in immunity to parasites. Nature 329:90–93

Lascano EF (1958) Mast cells in human tumors. Cancer 11:1110–1124

Lazar-Molnar E, Hegyesi H, Pallinger E et al (2002) Inhibition of human primary melanoma cell proliferation by histamine is enhanced by interleukin-6. Eur J Clin Invest 32:743–749

Leali D, Dell'Era P, Stabile H et al (2003) Osteopontin (Eta-1) and fibroblast growth factor-2 cross-talk in angiogenesis. J Immunol 171:1085–1093

Lebbink RJ, de Ruiter T, Adelmeijer J et al (2006) Collagens are functional, high affinity ligands for the inhibitory immune receptor LAIR-1. J Exp Med 203:1419–1425

Lee DM, Friend DS, Gurish MF et al (2002) Mast cells: a cellular link between autoantobodies and inflammatory arthritis. Science 297:1689–1692

Leek RD, Lewis CE, Whitehouse R et al (1996) Association of macrophage infiltration with angiogenesis and prognosis in invasive breast carcinoma. Cancer Res 56:4625–4629

Leek RD, Lander RJ, Harris AL et al (1999) Necrosis correlates with high vascular density and focal macrophages infiltration in invasive carcinoma of the breast. Br J Cancer 79:991–995

Lehtosalo JI, Uusitalo H, Laakso J et al (1984) Biochemical and immunohistochemical determination of 5-hydroxytryptamine located in mast cells in the trigeminal ganglion of the rat and guinea pig. Histochemistry 80:219–223

Leibovich SJ, Polverini PJ, Shetaprd HM et al (1987) Macrophage-induced angiogenesis is mediated by tumour necrosis factor-alpha. Nature 329:630–632

Lengauer C, Kinzler KW, Vogelstein B (1998) Genetic instabilities in human cancers. Nature 396:643–649

Lennert K, Parwaresch MR (1979) Mast cells and mast cell neoplasia; a review. Histopathology 3:349–365

Leulier F, Lemaitre B (2008) Toll-like receptors—taking an evolutionary approach. Nat Rev Genet 9:165–178

Levy AP, Levy NS, Wegner S et al (1995) Transcriptional regulation of the rat vascular endothelial growth factor gene by hypoxia. J Biol Chem 270:13333–13340

Lewis RA, Soter NA, Diamond PT et al (1982) Prostaglandin D2 generation after activation of rat and human mast cells with anti-IgE. J Immunol 129:1627–1631

Lewis CE, Leek R, Harris A et al (1995) Cytokine regulation of angiogenesis in breast cancer: the role of tumor-associated macrophages. J Leukoc Biol 57:747–751

Liebler JM, Picou MA, Qu Z et al (1997) Altered immunohistochemical localization of basic fibroblast growth factor after bleomycin-induced lung injury. Growth Factors 14:25–38

Lilla JN, Werb Z (2010) Mast cells contribute to the stromal microenvironment in mammary gland branching morphogenesis. Dev Biol 337:124–133

Lilla JN, Joshi RV, Craik CS et al (2009) Active plasma kallikrein localizes to mast cells and regulates epithelial cell apoptosis, adipocyte differentiation, and stromal remodeling during mammary gland involution. J Biol Chem 284:13792–13803

Lin EY, Pollard JW (2004) Role of infiltrated leucocytes in tumour growth and spread. Br J Cancer 90:2053–2058

Lin EY, Nguyen AV, Russel RG et al (2001) Colony-stimulating factor 1 promotes progression of mammary tumors to malignancy. J Exp Med 193:727–740

Lissbrant IF, Stattin P, Wikstrom P et al (2000) Tumor associated macrophages in human prostate cancer: relation to clinicopathological variables and survival. Int J Oncol 17:445–451

Liu J, Divoux A, Sun J et al (2009) Genetic deficiency and pharmacological stabilization of mast cells reduce diet-induced obesity and diabetics in mice. Nat Med 15:940–945
Lopez T, Hanahan D (2002) Elevated levels of IGF-1 receptor convey invasive and metastatic capability in a mouse model of pancreatic islet tumorigenesis. Cancer Cell 1:339–353
Lorentz A, Bischoff SC (2001) Regulation of human intestinal mast cells by stem cell factor and IL-4. Immunol Rev 179:57–60
Lorentz A, Schwengberg S, Sellge G et al (2000) Human intestinal mast cells are capable of producing different cytokine profiles: role of IgE receptor cross-linking and IL-4. J Immunol 164:43–48
Lorentz A, Schuppan D, Gebert A et al (2002) Regulatory effects of stem cell factor and interleukin-4 on adhesion of human mast cells to extracellular matrix proteins. Blood 99:966–972
Lorentz A, Wilke M, Sellge G et al (2005) IL-4-induced priming of human intestinal mast cells for enhanced survival and Th2 cytokine generation is reversible and associated with increased activity of ERK1/2 and c-Fos. J Immunol 174:6751–6756
Lu LF, Lind EF, Gondek DC et al (2006) Mast cells are essential intermediaries in regulatory T-cell tolerance. Nature 442:997–1002
Lundeberg L, El-Nour H, Mohabbati S et al (2002) Expression of serotonin receptors in allergic contact eczematous human skin. Arch Dermatol Res 294:393–398
Lyden D, Hattori K, Dias S et al (2001) Impaired recruitment of bone-marrow-derived endothelial and hematopoietic precursor cells blocks tumor angiogenesis and growth. Nat Med 7:1194–1201
Ma Z, Kwong KY, Tovar JP et al (2010) Cyclic adenosine monophosphate induces plasminogen activator inhibitor-1 expression in human mast cells. Biochem Biophys Res Commun 400:569–574
Maby-El Hajjami H, Petrova TV (2008) Developmental and pathological lymphangiogenesis: from models to human disease. Histochem Cell Biol 130:1063–1078
Machein MR, Knedla A, Knoth R et al (2004) Angiopoietin-1 promotes tumor angiogenesis in a rat glioma model. Am J Pathol 165:1557–1570
Mackins CJ, Kano S, Seyedi N et al (2006) Cardiac mast cell-derived renin promotes local angiotensin formation, norepinephrine release, and arrhythmias in ischemia/reperfusion J Clin Invest 116:1063–1070
Madden SL, Cook BP, Nacht M et al (2004) Vascular gene expression in nonneoplastic and malignant brain. Am J Pathol 165:601–608
Makitie T, Summanen P, Takkanen A et al (2001) Tumor-infiltrating macrophages (CD68(+) cells) and prognosis in malignant uveal melanoma. Invest Ophtalmol Vis Sci 42:1414–1421
Malaviya R, Ikeda T, Ross E et al (1996) Mast cell modulation of neutrophil influx and bacterial clearance at sites of infection through TNF-α. Nature 381:77–80
Mandriota SJ, Jussila L, Jeltsch M et al (2001) Vascular endothelial growth factor-C-mediated lymphangiogenesis promotes tumour growth. EMBO J 20:672–682
Maniotis AJ, Folberg R, Hess A et al (1999) Vascular channel formation by human melanoma cells in vivo and in vitro: vasculogenic mimicry. Am J Pathol 155:739–752
Mantovani A, Sozzani S, Locati M et al (2002) Macrophage polarization: tumor-associated macrophages as a paradigm for polarized M2 mononuclear phagocytes. Trends Immunol 23:549–555
Marks RM, Roche WR, Czerniecki M et al (1986) Mast cell granules cause proliferation of human microvascular endothelial cells. Lab Invest 55:289–294
Markwardt F, Klocking H (1977) Heparin-induced release of plasminogen activator. Haemostasis 6:370–374
Marone G, de Crescenzo G, Adt M et al (1995) Immunological characterization and functional importance of human heart mast cells. Immunopharmachology 31:1–18
Marone G, Florio G, Triggiani M et al (2000) Mechanisms of IgE elevation in HIV-1 infection. Crit Rev Immunol 20:477–496
Marshall JS (2004) Mast-cell responses to pathogens. Nat Rev Immunol 4:787–799
Martin R, Tam EK, Nadel JA et al (1992) Distribution of tryptase-containing mast cells in human bronchi. J Histochem Cytochem 40:781–786

Martínez L, Gimeno MF, Villán L et al (1999) Nitroprusside stimulates contractility and the synthesis of 14C-acetylated PAF-like substances in estrogen primed-mouse uterine horns. Prostaglandins Other Lipid Mediat 57:49–62
Masini E, Bechi P, Dei R et al (1994) Helicobacter pylori potentiates histamine release from rat serosal mast cells induced by bile acids. Dig Dis Sci 39:1493–1500
Massey WA, Guo CB, Dvorak AM et al (1991) Human uterine mast cells. Isolation, purification, characterization, ultrastructure, and pharmacology. J Immunol 147:1621–1627
Matsuda H, Kannan Y, Ushio H et al (1991) Nerve growth factor induces development of connective tissue-type mast cells in vitro from murine bone marrow cells. J Exp Med 174:7–14
Matsumoto I, Inoue Y, Shimada T et al (2001) Brain mast cells act as an immune gate to the hypothalamic-pituitary-adrenal axis in dogs. J Exp Med 194:71–78
Matsuzawa S, Sakashita K, Kinoshita T et al (2003) IL-9 enhances the growth of human mast cell progenitors under stimulation with stem cell factor. J Immunol 170:3461–3467
Masson V, De La Ballina LR, Munaut C et al. (2005) Contribution of host MMP-2 and MMP-9 to promote tumor vascularization and invasion of malignant keratinocytes. FASEB J 19:234–236
Maurer M, Echternacher B, Hültner L et al (1998) The c-kit ligand, stem-cell factor, can enhance innate immunity through effects on mast cells. J Exp Med 188:2343–2348
Maurer M, Theoharides T, Granstein RD et al (2003) What is the physiological function of mast cells? Exp Dermatol 12:886–910
Maurer M, Lopez Kostka S, Siebenhaar F et al (2006) Skin mast cells control T cell-dependent host defence in Leishmania major infections. FASEB J 20:2460–2467
McCurdy JD, Olynych TJ, Maher LH et al (2003) Cutting edge: distinct Toll-like receptor 2 activators selectively induces different classes of mediator production from human mast cells. J Immunol 170:1625–1629
McDonald DM, Boluk P (2002) Significance of blood vessel leakiness in cancer. Cancer Res 62:5381–5385
McDonald DM, Munn L, Jain RK (2000) Vasculogenic mimicry: how convincing, how novel, and how significant. Am J Pathol 156:383–388
McHardy P, Riley J, Huntley JF (1993) The recruitment of mast cells, exclusively of the mucosal phenotype, into granulomatous lesions caused by the pentastomid parasite Porocephalus crotali: recruitment is irrespective of site. Parasitology 106:47–54
McLachlan JB, Hart JP, Pizzo SV et al (2003) Mast cell-derived tumor necrosis factor induces hypertrophy of draining lymph nodes during infection. Nat Immunol 4:1199–1205
McLachlan JB, Shelburne CP, Hart JP et al (2008) Mast cell activators: a new class of highly effective vaccine adjuvants. Nat Med 14:536–541
McLaren KM, Pepper DS (1983) The immunoelectronmicroscopic localization of human platelet factor 4 in tissue mast cells. Histochem J 15:795–800
Meininger CJ, Yano H, Rottapel R et al (1992) The c-kit receptor ligand functions as a mast cell chemoattractant. Blood 79:958–963
Melillo RM, Guarino V, Avilla E et al (2010) Mast cells have a protumorigenic role in human thyroid cancer. Oncogene 29:6203–6215
Mellon MB, Frank BT, Fang KC (2002) Mast cell α-chymase reduces IgE recognition of birch pollen profiling by cleaving antibody-binding epitopes. J Immunol 168:290–297
Mellor EA, Maekawa A, Austen KF et al (2001) Cysteinyl leukotriene receptor 1 is also a pyrimidinergic receptor and is expressed by human mast cells. Proc Natl Acad Sci U S A 98:7964–7969
Merluzzi S, Frossi B, Gri G et al (2010) Mast cells enhance proliferation of B lymphocytes and drive their differentiation toward IgA-secreting plasma cells. Blood 115:2810–2817
Metcalfe DD (2008) Mast cells and mastocytosis. Blood 112:946–956
Metcalfe DD, Baram D, Mekori YA (1997) Mast cells. Physiol Rev 77:1033–1079
Metz M, Piliponsky AM, Chen CC et al (2006) Mast cells can enhance resistance to snake and honeybee venoms. Science 313:526–530
Metz M, Siebenhaar F, Maurer M (2008) Mast cell functions in the innate skin immune system. Immunobiology 213:251–269

Miller JS, Westin EH, Schwartz LB (1989) Cloning and characterization of complementary DNA for human tryptase. J Clin Invest 84:1188–1195
Miller JS, Moxley G, Schwartz LB (1990) Cloning and characterization of a second complementary DNA for human tryptase. J Clin Invest 86:864–870
Miller HR, Wright SH, Knight PA et al (1999) A novel function for transforming growth factor-β1: upregulation of the expression and the IgE-independent extracellular release of a mucosal mast cell granule-specific β-chymase, mouse mast cell protease-1. Blood 93:3473–3486
Mohammad-Zadeh LF, Moses L, Gwaltney-Brant SM (2008) Serotonin: a review. J Vet Pharmacol Ther 31:187–199
Molica S, Vacca A, Crivellato E et al (2003) Tryptase-positive mast cells predict clinical outcome of patients with early B-cell chronic lymphocytic leukemia. Eur J Haematol 71:137–139
Molica S, Montillo M, Ribatti D et al (2007) Intense reversal of bone marrow angiogenesis after sequential fludarabine-induction and alemtuzumab-consolidation therapy in advanced chronic lymphocytic leukemia. Haematologica 92:1367–1374
Molin D (2004) Bystander cells and prognosis in Hodgkin lymphoma. Review based on a doctoral thesis. Ups J Med Sci 109:179–228
Molin D, Edstrom A, Glimelius I et al (2002) Mast cell infilitration correlates with poor prognosis in Hodgkin's lymphoma. Br J Haematol 119:122–124
Moller A, Lippert U, Lessmann D et al (1993) Human mast cells produce IL-8. J Immunol 151:3261–3266
Morrey C, Brazin J, Seyedi N et al (2010) Interaction between sensory C-fibers and cardiac mast cells in ischemia/reperfusion: activation of a local renin-angiotensin system culminating in severe arrhythmic dysfunction. J Pharmacol Exp Ther 335:76–84
Mori R, Shaw TJ, Martin P (2008) Molecular mechanisms linking wound inflammation and fibrosis: knockdown of osteopontin leads to rapid repair and reduced scarring. J Exp Med 205:43–51
Morris BC, Danforth HD, Caldwell DJ et al (2004) Intestinal mucosal mast cell immune response and pathogenesis of two Eimeria acervulina isolates in broiler chickens. Poult Sci 83:1667–1674
Mota I, Vugman I (1956) Effects of anaphylactic shock and compound 48/80 on the mast cells of the guinea pig lung. Nature 177:427–429
Muhs BE, Plias G, Delgado Y et al (2003) Temporal expression and activation of matrix metalloproteinases-2, -9, and membrane type 1-matrix metalloproteinase following acute hindlimb ischemia. J Surg Res 111:8–15
Munn DH, Mellor AL (2006) The tumor-draining lymph node as an immune-privileged site. Immunol Rev 213:146–158
Muramatsu M, Katada J, Hattori M et al (2000a) Chymase mediates mast cell-induced angiogenesis in the hamster sponge granuloma. Eur J Pharmacol 402:181–191
Muramatsu M, Katada J, Hattori M et al (2000b) Chymase as a proangiogenic factor; a possible involvement of chymase-angiotensin-dependent pathway in the hamster sponge angiogenesis model. J Biol Chem 275:5545–5552
Murray DB, Gardner JD, Levick SP et al (2007) Response of cardiac mast cells to atrial natriuretic peptide. Am J Physiol Heart Circ Physiol 293:H1216–1222
Nakae S, Suto H, Hakurai M et al (2005) Mast cells enhance T cell activation: importance of mast cell-derived TNF. Proc Natl Acad Sci U S A 102:6467–6472
Nakae S, Suto H, Iikura M et al (2006) Mast cells enhance T cell activation: importance of mast cell costimulatory molecules and secreted TNF. J Immunol 176:2238–2248
Nakajima M, Irimura T, Di Ferrante N et al (1984) Metastatic melanoma heparanase. Characterization of heparan sulfate degradation fragments produced by B16 melanoma endoglucoronidase. J Biol Chem 259:2283–2290
Nakajima S, Krishnan B, Ota H et al (1997) Mast-cell involvement in gastritis with or without Helicobacter pylori infection. Gastroenterology 113:746–754
Nakao S, Kuwano T, Tsutsumi-Miyahara C et al (2005) Infiltration of COX-2-expressing macrophages is a prerequisite for IL-1 beta-induced neovascularization and tumor growth. J Clin Invest 115:2979–2991

Nakano T, Sonoda T, Hayashi C et al (1985) Fate of bone marrow-derived cultured mast cells after intracutaneous, intraperitoneal, and intravenous transfer into genetically mast cell-deficient W/Wv mice: evidence that cultured mast cells can give rise to both connective tissue type and mucosal mast cells. J Exp Med 162:1025–1043
Nakano N, Nishiyama C, Yagita H et al (2009) Notch signaling confers antigen-presenting cell functions on mast cells. J Allergy Clin Immunol 123:74–81
Nakayama T, Yao L, Tosato G (2004) Mast cell-derived angiopoietin-1 plays a critical role in the growth of plasma cell tumors. J Clin Invest 114:1317–1325
Nakayama T, Mutsuga N, Yao L et al (2006) Prostaglandin E2 promotes degranulation-independent release of MCP-1 from mast cells. J Leukoc Biol 79:95–104
Naldini A, Carraro F (2005) Role of inflammatory mediators in angiogenesis. Curr Drug Targets Inflamm Allergy 4:3–8
Naldini A, Leali D, Pucci A et al (2006) Cutting edge: IL-1beta mediates the proangiogenic activity of osteopontin-activated human monocytes. J Immunol 177:4267–4270
Nechushtan H (2010) The complexity of the complicity of mast cells in cancer. Int J Biochem Cell Biol 42:551–554
Nechushtan H, Razin E (2002) The function of MIFT and associated proteins in mast cells. Mol Immunol 38:1177–1180
Neri D, Caremolla B, Nissim A et al (1997) Targeting by affinity-matured recombinant antibody fragments of an angiogenesis associated fibronectin isoform. Nat Biotechnol 15:1271–1275
Nickel R, Beck LA, Stellato C, Schleimer RP (1999) Chemokines and allergic disease. J Allergy Clin Immunol 104:723–742
Nico B, Mangieri D, Benagiano V et al (2008a) Nerve growth factor as an angiogenic factor. Microvasc Res 75:135–141
Nico B, Mangieri D, Crivellato E et al (2008b) Mast cells contribute to vasculogenic mimicry in multiple myeloma. Stem Cells Dev 17:19–22
Nicolson GL (1988a) Cancer metastasis: tumor cell and host organ properties important in metastasis to specific secondary sites. Biochim Biophys Acta 948:175–224
Nicolson GL (1988b) Organ specificity of tumor metastasis: role of preferential adhesion, invasion and growth of malignant cells at specific secondary sites. Cancer Metastasis Rev 7:143–188
Nicolson GL, Dulski KM (1986) Organ specificity of metastatic tumor colonization is related to organ-selective growth properties of malignant cells. Int J Cancer 38:289–294
Nielsen HJ, Christenseb IJ, Svendsen MN et al (2002) Ranitidine as adjuvant treatment in colorectal cancer. Br J Surg 89:1416–1422
Nigrovic PA, Lee DM (2007) Synovial mast cells: role in acute and chronic arthritis. Immunol Rev 217:19–37
Niimi Y, Pawankar R, Kawana S (2006) Increased expression of matrix metalloproteinase-2, matrix metalloproteinase-9 and matrix metalloproteinase-13 in lesional skin of bullous pemphigoid. Int Arch Allergy Immunol 139:104–113
Nilsson G, Forsberg-Nilsson K, Xiang Z, Hallbook F et al (1997) Human mast cells express functional TrkA and are a source of nerve growth factor. Eur J Immunol 27:2295–2301
Nishie A, Ono M, Shono T et al (1999) Macrophage infiltration and heme oxygenase-1 expression correlate with angiogenesis in human gliomas. Clin Cancer Res 5:1107–1113
Nishiyama C, Ito T, Nishiyama M et al (2005) GATA-1 is required for expression of Fc{epsilon} RI on mast cells: analysis of mast cells derived from GATA-1 knockdown mouse bone marrow. Int Immunol 17:847–856
Niyonsaba F, Ushio H, Hara M et al (2010) Antimicrobial peptides human beta-defensins and cathelicidin LL-37 induce the secretion of a pruritogenic cytokine IL-31 by human mast cells. J Immunol 184:3526–3534
Niwa Y, Kasugai T, Ohno K et al (1991) Anemia and mast cells depletion in mutant rats that are homozygous at 'white spotting (Ws)' locus. Blood 78:1936–1941
Nonomura N, Takayama H, Nishimura K et al (2007) Decreased number of mast cells infiltrating into needle biopsy specimens leads to a better prognosis of prostate cancer. Br J Cancer 97:952–956

Norrby K (1993) Heparin and angiogenesis: a low molecular weight fraction inhibits and a high molecular weight fraction stimulates angiogenesis systematically. Haemostasis 23:144–149
Norrby K, Sorbo J (1992) Heparin enhances angiogenesis by a systemic mode of action. Int J Exp Pathol 73:1451–1455
Norrby A, Jakobsson A, Sorbo J (1986) Mast cell-mediated angiogenesis: a novel experimental model using the rat mesentery. Virchows Arch B Cell Pathol Incl Mol Pathol 52:195–206
Norris AA (1996) Pharmacology of sodium cromoglycate. Clin Exp Allergy 26 (S4):5–7
Nozawa H, Chiu C, Hanahan D (2006) Infiltrating neutrophils mediate the initial angiogenic switch in a mouse model of multistage carcinogenesis. Proc Natl Acad Sci USA 103:12493–12498
Obermüller E, Vosseler S, Fusenig NE et al (2004) Cooperative autocrine and paracrine functions of granulocyte colony-stimulating factor and granulocyte-macrophage colony-stimulating factor in the progression of skin carcinoma cells. Cancer Res 64:7801–7812
Ochi H, Hirani WM, Yuan Q et al (1999) T helper cell type 2 cytokine-mediated comitogenic responses and CCR3 expression during differentiation of human mast cells in vitro. J Exp Med 190:267–280
Oguma T, Palmer LJ, Birben E et al (2004) Role of prostanoid DP receptor variants in susceptibility to asthma. N Engl J Med 351:1752–1763
Oh SW, Pae CI, Lee DK et al (2002) Tryptase inhibition blocks airway inflammation in a mouse asthma model. J Immunol 168:1992–2000
Ohno S, Inagawa H, Soma G et al (2002) Role of tumor-associated macrophage in malignant tumors: should the location of the infiltrated macrophages be taken into account during evaluation? Anticancer Res 22:4269–4275
Ohno S, Ohno Y, Suzuki N et al (2004) Correlation of histological localization of tumor-associated macrophages with clinicopathological features in endometrial cancer. Anticancer Res 24:3335–3342
Okamura K, Sato Y, Matsuda T et al (1991) Endogenous basic fibroblast growth fcator-dependent induction of collagenase and interleukin-6 in tumor necrosis factor-treated human microvascular endothelial cells. J Biol Chem 266:19162–19165
Okayama Y, Kawakami T (2006) Development, migration, and survival of mast cells. Immunol Res 34:97–115
Okayama Y, Kirshenbaum AS, Metcalfe DD (2000) Expression of a functional high-affinity IgG receptor, Fc gamma RI, on human mast cells: up-regulation by IFN-gamma. J Immunol 164:4332–4339
Okayama Y, Hagaman DD, Metcalfe DD (2001) A comparison of mediators released or generated by IFN-gamma-treated human mast cells following aggregation of Fc gamma RI or Fc epsilon RI. J Immunol 166:4705–4712
Okayama Y, Ra C, Saito H (2007) Role of mast cells in airway remodeling. Curr Opin Immunol 19:687–693
Okayama Y, Okumura S, Sagara H et al (2009) FcepsilonRI-mediated thymic stromal lymphopoietin production by interleukin-4-primed human mast cells. Eur Respir J 34:425–435
Okumura S, Sagara H, Fukuda T et al (2005) FcepsilonRI-mediated amphiregulin production by human mast cells increases mucin gene expression in epithelial cells. J Allergy Clin Immunol 115:272–279
Oliveira-Neto HH, Ferreira Leite A, Lago Costa N et al (2007) Decrease in mast cells in oral squamous cell carcinoma: possible failure in the migration of these cells. Oral Oncol 43:484–490
Olsson N, Ulfgren AK, Nilsson G (2001) Demonstration of mast cell chemotactic activity in synovial fluid from rheumatoid patients. Ann Rheum Dis 60:187–193
Orange RP, Austen WG, Austen KF (1971) Immunological release of histamine and slow-reacting substance of anaphylaxis from human lung. I. Modulation by agents influencing cellular levels of cyclic 3′,5′-adenosine monophosphate. J Exp Med 134:136s–148s
Orimo A, Gupta PB, Sgroi DC et al (2005) Stromal fibroblasts present in invasive human breast carcinomas promote tumor growth and angiogenesis through elevated SDF-1/CXCL12 secretion. Cell 121:335–348

Orinska Z, Bulanova E, Budagian V et al (2005) TLR3-induced activation of mast cells modulates CD8+T-cell recruitment. Blood 106:978–987
Orinska Z, Maurer M, Mirghomizadeh F et al (2007) IL-15 constrains mast cell-dependent antibacterial defences by suppressing chymase activities. Nature Med 13:927–934
Ott VL, Cambier JC, Kappler J et al (2003) Mast cell-dependent migration of effector CD8+ T cells through production of leukotriene B4. Nat Immunol 4:974–981
Papetti M, Herman IM (2002) Mechanisms of normal and tumor-derived angiogenesis. Am J Physiol Cell Physiol 282:C947–C970
Pardanani A, Elliott M, Reeder T, Li CY et al (2003) Imatinib for systemic mast cell disease. Lancet 362:535–536
Pardoll D (2003) Does the immune system see tumors as foreign or self? Annu Rev Immunol 21:807–839
Park CC, Bissell MJ, Barcellos-Hoff MH (2000) The influence of the microenvironment on the malignant phenotype. Mol Med Today 6:324–329
Parker BS, Argani P, Cook BP et al (2004) Alterations in vascular gene expression in invasive breast carcinoma. Cancer Res 64:7857–7866
Parshad R, Kapoor S, Gupta SD et al (2002) Does famotidine enhance tumor infilitrating lymphocytes in breast cancer? Results of a randomized prospective pilot study. Acta Oncol 41:362–365
Parratt JR, West GB (1956) The location and possible function of tissue 5-hydroxytryptamine in the rat. J Physiol 134:11P–12P
Pasqualini R, Arap W (2002) Translation of vascular diversity into targeted therapeutics. Ann Hematol 81(Suppl 2):S66–S67
Patella V, Marinò I, Lampärter B et al (1995) Human heart mast cells. Isolation, purification, ultrastructure, and immunologic characterization. J Immunol 154:2855–2865
de Paulis A, Marinò I, Ciccarelli A et al (1996) Human synovial mast cells. I. Ultrastructural in situ and in vitro immunologic characterization. Arthritis Rheum 39:1222–1233
Peng SH, Deng H, Yang JF et al (2005) Significance and relationship between infiltrating inflammatory cells and tumor angiogenesis in hepatocellular carcinoma tissues. World J Gastroenterol 11:6521–6523
Pezzella F, Pastorino U, Tagliabue E et al (1997) Non-small-lung carcinoma tumor growth without morphological evidence of neo-angiogenesis. Am J Pathol 151:1417–1423
Piconese S, Gri G, Tripodo C et al (2009) Mast cells counteract regulatory T-cell suppression through interleukin-6 and OX40/OX40L axis toward Th17-cell differentiation. Blood 114:2639–2648
Piliponsky AM, Chen CC, Nishimura T et al (2008) Neurotensin increases mortality and mast cells reduce neurotensin levels in a mouse model of sepsis. Nat Med 14:392–398
Pilgrim HI (1969) The kinetics of the organ-specific metastasis of a transplantable reticuloendothelial tumor. Cancer Res 29:1200–1205
Plate KH, Breier G, Widch HA et al (1992) Vascular endothelial growth factor is a potent tumour angiogenesis factor in human gliomas in vivo. Nature 359:845–848
Plaut M, Pierce JH, Watson CJ et al (1989) Mast cell lines produce lymphokines in response to cross-linkage of FcεRI or to calcium ionophores. Nature 339:64–67
Plebani M, Basso D, Vianello F et al (1994) Helicobacter pylori activates gastric mucosal mast cells. Dig Dis Sci 39:1592–1593
Poglio S, De Toni-Costes F, Arnaud E et al (2010) Adipose tissue as a dedicated reservoir of functional mast cell progenitors. Stem Cells 28:2065–2072
Poole TJ, Zetter BR (1983) Stimulation of rat peritoneal mast cell migration by tumor-derived peptides. Cancer Res 43:5857–5861
Potgens AJ, Lubsen NH, van Altena MC et al (1995) Vascular permeability factor expression influences tumor angiogenesis in human melanoma lines xenografted to nude mice. Am J Pathol 146:197–209
Prausnitz C, Küstner H (1921) Studien über die Überempfindlichkeit. Zentralbl Bakteriol Parasitenk Infektionsk Abt 1 Orig 86:160–169

Prodeus AP, Zhou X, Maurer M (1997) Impaired mast cell-dependent natural immunity in complement C3-deficient mice. Nature 390:172–175
Pulimood AB, Mathan MM, Mathan VI (1998) Quantitative and ultrastructural analysis of rectal mucosal mast cells in acute infectious diarrhea. Dig Dis Sci 43:2111–2116
Puxeddu I, Ribatti D, Crivellato E et al (2005) Mast cells and eosinophils: a novel link between inflammation and angiogenesis in allergic diseases. J Allergy Clin Immunol 116:531–536
Qu Z, Kayton RJ, Ahmadi P et al (1998) Ultrastructural immunolocalization of basic fibroblast growth factor in mast cell secretory granules: morphological evidence for bFGF release through degranulation. J Histochem Cytochem 46:1119–1128
Queen MM, Ryan RE, Holzer RG et al (2005) Breast cancer cells stimulate neutrophils to produce oncostatin M: potential implications for tumor progression. Cancer Res 65:8896–8904
Quenslen U (1933) Studien über die Gewebsmastzellen. Acta Path Micorbiol Scand Suppl 16:358–375
Quiao H, Andrade MV, Lisboa FA et al (2006) FcepsilonR1 and toll-like receptors mediate synergistic signals to markedly augment production of inflammatory cytokines in murine mast cells. Blood 107:610–618
Ra C, Yasuda M, Yagita H, Okumura K (1994) Fibronectin receptor integrins are involved in mast cell activation. J Allergy Clin Immunol 99:625–628
Rak J, Mitsuhashi Y, Bayko L et al (1995) Mutant ras oncogenes upregulate VEGF/VPF expression: implications for induction and inhibition of tumor angiogenesis. Cancer Res 55:4575–4580
Ramakrishnan G, Jagan S, Kamaraj S et al (2009) Silymarin attenuated mast cell recruitment thereby decreased the expressions of matrix metalloproteinases-2 and 9 in rat liver carcinogenesis. Invest New Drugs 27:233–240
Rapport M, Green A, Page I (1948) Crystalline serotonin. Science 108:329–330
Rastinejad F, Polverini PJ, Bouck NP (1989) Regulation of the activity of a new inhibitor of angiogenesis by a cancer suppressor gene. Cell 56:345–355
Rauter I, Krauth MT, Westritschnig K et al (2008) Mast cell-derived proteases control allergic inflammation through cleavage of IgE. J Allergy Clin Immunol 121:197–202
Raymond WW, Waugh Ruggles S, Craik CS et al (2003) Albumin is a substrate of human chymase: prediction by combinatorial peptide screening and development of a selective inhibitor based on the albumin cleavage site. J Biol Chem 278:34517–34524
Raza SL, Cornelius LA (2000) Matrix metalloproteinases: pro- and anti-angiogenic activities. J Invest Dermatol Symp Proc 5:47–54
Reading CL, Hutchins JF (1985) Carbohydrate structure in tumor immunity. Cancer Metastasis Rev 4:221–260
Reilly CF, Tewksbury DA, Schechter NB et al (1982) Rapid conversion of angiotensin I to angiotensin II by neutrophil and mast cell proteinases. J Biol Chem 257:8619–8622
Reinheimer T, Möhlig T, Zimmermann S et al (2000) Muscarinic control of histamine release from airways. Inhibitory M1-receptors in human bronchi but absence in rat trachea. Am J Respr Crit Care Med 162:534–538
Regaud C, Lacassagne A (1922) A propos des mastocytes des épithélomas. Importance de la fixation pour la colouration des granulations des mastocytes. C R Soc Biol (Paris) 87:1084–1086
Ribatti D, Crivellato E (2009) Immune cells and angiogenesis. J Cell Mol Med 13:2822–2833
Ribatti D, Ranieri G, Nico B, et al (2011) Tryptase and chymase are angiogenic *in vivo* in the chorioallantoic membrane assay. Int J Dev Biol 55:99–102
Ribatti D, Roncali L, Nico B et al (1987) Effects of exogenous heparin on the vasculogenesis of the chorioallantoic membrane. Acta Anat 130:257–263
Ribatti D, Nico B, Vacca A et al (1998) Do mast cells help to induce angiogenesis in B-cell non-Hodgkin's lymphomas? Br J Cancer 77:1900–1906
Ribatti D, Vacca A, Dammacco F (1999a) The role of the vascular phase in solid tumor growth: a historical review. Neoplasia 1:293–302
Ribatti D, Vacca A, Nico B et al (1999b) Bone marrow angiogenesis and mast cell density increase simultaneously with progression of human multiple myeloma. Br J Cancer 79:451–455

Ribatti D, Vacca A, Marzullo A et al (2000) Angiogenesis and mast cell density with tryptase activity increase simultaneously with pathological progression in B-cell non-Hodgkin's lymphomas. Int J Cancer 82:171–175
Ribatti D, Crivellato E, Candussio L et al (2001a) Mast cells and their secretory granules are angiogenic in the chick embryo chorioallantoic membrane. Clin Exp Allergy 31:602–608
Ribatti D, Vacca A, Nico B et al (2001b) The role of mast cells in tumour angiogenesis. Br J Haematol 115:514–521
Ribatti D, Polimeno G, Vacca A et al (2002) Correlation of bone marrow angiogenesis and mast cells with tryptase activity in myelodysplastic syndromes. Leukemia 16:1680–1684
Ribatti D, Ennas MG, Vacca A et al (2003a) Tumor vascularity and tryptase positive-mast cells correlate with a poor prognosis in melanoma. Eur J Clin Invest 33:420–425
Ribatti D, Vacca A, Ria R et al (2003b) Neovascularization, expression of fibroblast growth factor-2, and mast cell with tryptase activity increase simultaneously with pathological progression in human malignant melanoma. Eur J Cancer 39:666–675
Ribatti D Molica S, Vacca A et al (2003c) Tryptase-positive mast cells correlate positively with bone marrow angiogenesis in B-cell chronic lymphocytic leukemia. Leukemia 17:1428–1430
Ribatti D, Nico B, Spinazzi R et al (2005a) The role of adrenomedullin in angiogenesis. Peptides 26:1670–1675
Ribatti D, Finato N, Crivellato E et al (2005b) Neovascularization and mast cells with tryptase activity increase simultaneously with pathologic progression in human endometrial cancer. Am J Obstet Gynecol 193:1961–1965
Ribatti D, Mangialardi G, Vacca A (2006) Stephen Paget and the 'seed and soil' theory of metastatic dissemination. Clin Exp Med 6:145–149
Ribatti D, Belloni AS, Nico B et al (2007a) Tryptase- and leptin-positive mast cells correlate with vascular density in uterine leiomyomas. Am J Obstet Gynecol 196:470.e1–470.e7
Ribatti D, Nico B, Crivellato E et al (2007b) The history of the angiogenic switch concept. Leukemia 21:44–52
Ribatti D, Nico B, Crivellato E et al (2007c) The structure of the vascular networks of tumors. Cancer Letters 248:18–23
Ribatti D, Finato N, Crivellato E et al (2007d) Angiogenesis and mast cells in human breast cancer sentinel lymph nodes with and without micrometastases. Histopathology 51:837–842
Ribatti D, Nico B, Finato N et al (2009) Co-localization of tryptase and cathepsin-G in mast cells in cutaneous mastocytosis. Cancer Letters 279:209–212
Ribatti D, Guidolin D, Marzullo A et al (2010) Mast cells and angiogenesis in gastric carcinoma. Int J Exp Pathol 91:350–356
Ribatti D, Ranieri G, Nico B et al (2011) Tryptase and chymase are angiogenic in vivo in the CAM assay. Int J Dev Biol 55:99–102
Rihet P, Demeure CE, Bourgois A et al (1991) Evidence for an association between human resistance to Schistosoma mansoni and high anti-larval IgE levels. Eur J Immunol 21:2679–2686
Riley JF, West GB (1952) Histamine in tissue mast cells. J Physiol 117:72P–73P
Riley JF (1954) Heparin, histamine and mast cells. Blood 9:1123–1126
Riley JF (1955) Pharmacology and functions of mast cells. Pharmacol Rev 7:267–277
Riley JF (1963) Functional significance of histamine and heparin in tissue mast cells. Ann N Y Acad Sci 103:151–163
Riley JF, West GB (1953) The presence of histamine in tissue mast cells. J Physiol 120:528–537
Ringvall M, Rönnberg E, Wernersson S et al (2008) Serotonin and histamine storage in mast cell secretory granules is dependent on serglycin proteoglycan. J Allergy Clin Immunol 121:1020–1026
Ritter U, Meissner A, Ott J et al (2003) Analysis of the maturation process of dendritic cells deficient for TNF and lymphotoxin-α reveals an essential role for TNF. J Leukoc Biol 74:216–222
Rivera J, Gilfillan AM (2006) Molecular regulation of mast cell activation. J Allergy Clin Immunol 117:1214–1225

Roberts LJ, Lewis RA, Oates JA et al (1979) Prostaglandin tromboxane, and 12-hydroxy-5,8,10,14-eicosatetraenoic acid production by ionophore-stimulated rat serosal mast cells. Biochim Biophys Acta 575:185–192
Rocha E, Silva M (1952) Concerning the mechanism of anaphylaxis and allergy. Brit Med J 1:779–784
Roche WR (1986) The nature and significance of tumour-assisted mast cells. J Pathol 148:175–182
Rodewald HR, Dessing M, Dvorak AM et al (1996) Identification of a committed precursor for the mast cell lineage. Science 271:818–822
Rojas IG, Spencer ML, Martinez A et al (2005) Characterization of mast cell subpopulations in lip cancer. J Oral Pathol Med 34:268–273
Rojas IG, Martinez A, Brethauer U et al (2009) Actinic cheilitis: epithelial expression of COX-2 and its association with mast cell tryptase and PAR-2. Oral Oncol 45:284–290
Romagnani P, De Paulis A, Beltrame C et al (1999) Tryptase-chymase double positive human mast cells express the eotaxin receptor (CCR3) and are attracted by CCR3-binding chemokines. Am J Pathol 155:1195–1204
Romagnani P, Annunziato F, Lasagni L et al (2001) Cell cycle-dependent expression of CXC chemokine receptor 3 by endothelial cells mediates angiostatic activity. J Clin Invest 107:53–63
Rose ME, Ogilvie BM, Bradley JWA (1980) Intestinal mast cell response in rats and chickens to coccidiosis, with some properties of chicken mast cells. Int Arch Allergy Appl Immunol 63:21–29
Rosenkranz AR, Coxon A, Maurer M et al (1998) Impaired mast cell development and innate immunity in Mac-1 (CD11b/CD18, CR3)-deficient mice. J Immunol 161:6463–6467
Rudolph MI, de los Angeles García M, Sepulveda M et al (1997) Ethodin: pharmacological evidence of the interaction between smooth muscle and mast cells in the myometrium. J Pharmacol Exp Ther 282:256–261
Ruoslahti E (2002) Specialization of tumour vasculature. Nat Rev Cancer 2:83–90
Ruoss SJ, Hartmann T, Caughey GH (1991) Mast cell tryptase is a mitogen for cultured fibroblasts. J Clin Invest 88:493–499
Rychter JW, Van Nassauw L, Timmermans JP et al (2011) CGRP1 receptor activation induces piecemeal release of protease-1 from mouse bone marrow-derived mucosal mast cells. Neurogastroenterol Motil 23:e57–68
Saito H (2005) Culture of human mast cells from hemopoietic progenitors. Methods Mol Biol 315:113–122
Sakaguchi M, Takai S, Jin D et al (2004) A specific chymase inhibitor, NK3201, suppresses bleomycin-induced pulmonary fibrosis in hamsters. Eur J Pharmacol 493:173–176
Sakamoto S, Goldhaber P, Glimcher M (1973) The effect of heparin on the amount of enzyme released in tissue culture and on the activity of the enzyme. Calcif Tissue Int 12:247–258
Sakata-Yanagimoto M, Sakai T, Miyake Y et al (2011) Notch2 signaling is required for proper mast cell distribution and mucosal immunity in the intestine. Blood 117:128–134
Samoszuk M, Corvin MA (2003a) Acceleration of tumor growth and peri-tumoral blood clotting by imatinib mesylate (Gleevec). Int J Cancer 106:647–652
Samoszuk M, Corwin MA (2003b) Mast cell inhibitor cromolyn increases blood clotting and hypoxia in murine breast cancer. Int J Cancer 107:159–163
Sasaki M, Kashima M, Ito T et al (2000) Effect of heparin and related glycosaminoglycan on PDGF-induced lung fibroblast proliferation, chemotactic response and matrix metalloproteinases activity. Mediators Inflamm 9:85–91
Sawatsubashi M, Yamada T, Fukushima N et al (2000) Association of vascular endothelial growth factor and mast cells with angiogenesis in laryngeal squamous cell carcinoma. Virchows Arch 436:243–248
Sayed BA, Brown MA (2007) Mast cells as modulators of T-cell responses. Immunol Rev 217:53–64
Sayed BA, Christy A, Quirion MR (2008) The master switch: the role of mast cells in autoimmunity and tolerance. Annu Rev Immunol 26:705–739

Schadendorf D, Moller A, Algermissen B et al (1993) IL-8 produced by human malignant melanoma cells in vitro is an essential autocrine growth factor. J Immunol 151:2667
Schechter NM, Choi JK, Slavin DA et al (1986) Identification of a chymotrypsin-like proteinase in human mast cells. J Immunol 137:962–970
Schiemann F, Brandt E, Gross R (2009) The cathelicidin LL-37 activates human mast cells and is degraded by mast cell tryptase: counter-regulation by CXCL4. J Immunol 183:2223–2231
Schilder RJ, Sill MW, Lee RB (2008) Phase II evaluation of imatinib mesylate in the treatment of recurrent or persistent epithelial ovarian or primary peritoneal carcinoma: a Gynecologic Oncology Group Study. J Clin Oncol 26:3418–3426
Schmeisser A, Garlichs CD, Zhang H et al (2001) Monocytes coexpress endothelial and macrophagocytic lineage markers and form cord-like structures in Matrigel under angiogenic conditions. Cardiovasc Res 49:671–680
Schwandner R, Dziarski R, Wesche H et al (1999) Peptidoglycan- and lipotheichoic acid-induced cell activation is mediated by toll-like receptor 2. J Biol Chem 274:17406–17409
Schwartz LB, Lewis RA, Austen KF (1981) Tryptase from human pulmonary mast cells: purification and characterization. J Biol Chem 256:11939–11943
Schwartz LB, Metcalfe DD, Miller JS et al (1987a) Tryptase levels as an indicator of mast-cell activation in systemic anaphylaxis and mastocytosis. N Engl J Med 316:1622–1626
Schwartz LB, Irani AA, Roller K et al (1987b) Quantitation of histamine, tryptase and chymase in dispersed human T and TC mast cells. J Immunol 138:2611–2615
Schwartz JD, Monea S, Marcus SG et al (1998) Soluble factor(s) released from neutrophils activates endothelial cell matrix metalloproteinase-2. J Surg Res 76:79–85
Seftor RB, Seftor EA, Koshikawa N et al (2001) Cooperative interactions of laminin 5 gamma-2 chain, matrix metalloproteinase-2, and membrane type-1-matrix/metalloproteinase are required for mimicry of embryonic vasculogenesis by aggressive melanoma. Cancer Res 61:6322–6327
Seibold JR, Giorno RC, Claman HN (1990) Dermal mast cell degranulation in systemic sclerosis. Arthritis Rheum 33:1702–1709
Semenza GL (1996) Transcriptional regulation by hypoxia-inducible factor-1. Trends Cardiovasc Med 6:151–157
Shamamian P, Schwartz JD, Pocock BJ et al (2001) Activation of progelatinase A (MMP-2) by neutrophil elastase, cathepsin G, and proteinase-3: a role for inflammatory cells in tumor invasion and angiogenesis. J Cell Physiol 189:197–206
Sharma VK, Agrawal AK, Pratarp VK et al (1992) Mast cell reactivity in lymphoma: a preliminary communication. India J Cancer 29:61–65
Sharma B, Sriram G, Saraswathi TR et al (2010) Immunohistochemical evaluation of mast cells and angiogenesis in oral squamous cell carcinoma. Indian J Dent Res, 21:260–265
Sher A, Hein A, Moser G et al (1979) Complement receptors promote the phagocytosis of bacteria by rat peritoneal mast cells. Lab Invest 41:490–499
Shiang C, Mauad T, Senhorini A et al (2009) Pulmonary periarterial inflammation in fatal asthma. Clin Exp Allergy 39:1450–1452
Shin JS, Gao Z, Abraham SN (2000) Involvement of cellular caveolae in bacterial entry into mast cells. Science 289:785–788
Shin JW, Min M, Larrieu-Lahargue F et al (2006) Prox1 promotes lineage-specific expression of fibroblast growth factor (FGF) receptor-3 in lymphatic endothelium: a role for FGF signaling in lymphangiogenesis. Mol Biol Cell 17:576–584
Shin K, Watts GF, Oettgen HC et al (2008) Mouse mast cell tryptase mMCP-6 is a critical link between adaptive and innate immunity in the chronic phase of Trichinella spiralis infection. J Immunol 180:4885–4891
Shweiki D, Itin A, Soffer D et al (1992) Vascular endothelial growth factor induced by hypoxia may mediate hypoxia-initiated angiogenesis. Nature 359:843–845
Sica A, Schioppa T, Mantovani A et al (2006) Tumour-associated macrophages are a distinct M2 polarised population promoting tumour progression: potential targets of anti-cancer therapy. Eur J Cancer 42:717–727

Siddiqui EJ, Thompson CS, Mikhailidis DP et al (2005) The role of serotonin in tumour growth (review). Onc Rep 14:1593–1597

Sillaber C, Baghestanian M, Hofbauer R et al (1997) Molecular and functional characterization of the urokinase receptor on human mast cells. J Biol Chem 272:7824–7832

Slominski A, Pisarchik A, Johansson O et al (2003) Tryptophan hydroxylase expression in human skin cells. Biochim Biophys Acta 1639:80–86

Spitz DR (2000) Glucose deprivation-induced oxidative stress in human tumor cells. Ann N Y Acad Sci 899:349–362

Spurbeck WW, NG CY, Strom TS et al (2002) Enforced expression of tissue inhibitor of matrix metalloproteinase-3 affects functional capillary morphogenesis and inhibits tumor growth in a murine tumor model. Blood 2002; 100:3361–3368

Sorbo J, Jakobson A, Norrby K (1994) Mast cell histamine is angiogenic through receptors for histamine 1 and histamine 2. Int J Exp Pathol 75:43–50

Soucek L, Lawlor ER, Soto D et al (2007) Mast cells are required for angiogenesis and macroscopic expansion of Myc-induced pancreatic islet tumors. Nat Med 13:1211–1218

Stack MS, Johnson DA (1994) Human mast cell tryptase activates single-chain urinary-type plasminogen activator (pro-urokinase). J Biol Chem 269:9416–9419

Staemmler M (1921) Untersuchung über Vorkommen und Bedeutung der histiogenen Mastzellen im menschlichen Körper unter normalen und pathologischen Verhältnissen. Frankfurt Z Path 25:391–435

Starkey JR, Crowle PK, Taubenberger S (1988) Mast cell-deficient W/Wv mice exhibit a decreased rate of tumor angiogenesis. Int J Cancer 42:48–52

St Croix B, Rago C, Velculescu V et al (2000) Genes expressed in human tumor endothelium. Science 289:1197–1202

Stead RH, Tomioka M, Quinonez G et al (1987) Intestinal mucosal mast cells in normal and nematode-infected rat intestines are in intimate contact with peptidergic nerves. Proc Natl Acad Sci U S A 84:2975–2987

Stead RH, Dixon MF, Bramwell NH et al (1989) Mast cells are closely apposed to nerves in the human gastrointestinal mucosa. Gastroenterology 97:575–585

Stead RH, Kosecka-Janiszewska U, Oestreicher AB et al (1991) Remodeling of B-50 (GAP-43)- and NSE-immunoreactive mucosal nerves in the intestines of rats infected with Nippostrongylis brasiliensis. J Neurosci 11:3809–3821

Stelekati E, Orinska Z, Bulfone-Paus S (2007) Mast cells in allergy: innate instructors of adaptive responses. Immunobiology 212:505–519

Sterinberg F, Rohrborn HJ, Otto T et al (1997) NIR reflection measurements of hemoglobin and cytochrome da3 in healthy tissue and tumors. Correlation to oxygen consuption: preclinical and clinical data. Adv Exp Med Biol 428:69–77

Stoeltzing O, Ahmad SA, Liu W et al (2003) Angiopoietin-1 inhibits vascular permeability, angiogenesis, and growth of hepatic colon cancer tumors. Cancer Res 63:3370–3377

Stromblad S, Cheresh DA (1996) Cell adhesion and angiogenesis. Trends Cell Biol 6:462–468

Suekane T, Ikura Y, Watanabe K et al (2010) Phenotypic change and accumulation of smooth muscle cells in structures in Crohn's disease: relevance to local angiotensin II system. J Gastroenterol 45:821–830

Sugiyama D, Tanaka M, Kitajima K et al (2008) Differential context-dependent effects of friend of GATA-1 (FOG-1) on mast-cell development and differentiation. Blood 111:1924–1932

Sun J, Sukhova GK, Wolters et al (2007a) Mast cells promote atherosclerosis by releasing proinflammatory cytokines. Nat Med 13:719–724

Sun J, Sukhova GK, Yang M et al (2007b) Mast cells modulate the pathogenesis of elastase-induced abdominal aortic aneurysms in mice. J Clin Invest 117:3359–3368

Sunderkokker C, Goebeler M, Schiltze-Osthoff K et al (1991) Macrophage-derived angiogenesis factors. Pharmacol Ther 51:195–216

Suto H, Nakae S, Kakurai M et al (2006) Mast cell-associated TNF promotes dendritic cell migration. J Immunol 176:4102–4112

Sutton BJ, Gould HJ (1993) The human IgE network. Nature 366:421–428

Sylvèn B (1940) Studies on the liberation of sulphuric acids from the granules of mast cells in the subcutaneous connective tissue after exposure to Roengten and gamma rays. Acta Radiol (Stockh) 21:206–212

Syväranta S, Helske S, Laine M et al (2010) Vascular endothelial growth factor-secreting mast cells and myofibroblasts: a novel self-perpetuating angiogenic pathway in aortic valve stenosis. Arterioscler Thromb Vasc Biol 30:1220–1227

Taghon T, Yui MA, Rothenberg EV (2007) Mast cell lineage diversion of T lineage precursors by the essential T cell transcription factor GATA-3. Nat Immunol 8:845–855

Taipale J, Lohi, J, Saarinen J et al (1995) Human mast cell chymase and leukocyte elestase release latent transforming growth factor beta-1 from the extracellular matrix of cultured human epithelial and endothelial cells. J Biol Chem 270:4689–4696

Takahashi T, Kalka C, Masuda H et al (1999) Ischemia- and cytokine-induced mobilization of bone marrow-derived endothelial progenitor cells for neovascularization. Nat Med 5:434–438

Takanami I, Takeuchi K, Narume M (2000) Mast cell density is associated with angiogenesis and poor prognosis in pulmonary adenocarcinoma. Cancer 88:2686–2692

Talkington J, Nickell SP (2001) Role of Fgγ receptors in triggering host-cell activation and cytokine release by Borrelia burgdorferi. Infect Immunol 69:413–419

Talmadge JE, Fidler IJ (1982) Cancer metastasis is selective or random depending on the parent tumour population. Nature 297:593–594

Tanaka A, Matsuda H (2004) IgE crosslinkage of Fcepsilon receptor I induces both production and activation of matrix metalloproteinase-9 in mast cells. Cell Immunol 228:66–75

Tanaka A, Yamane Y, Matsuda H (2001) Mast cell MMP-9 production enhanced by bacterial lipopolysaccharide. J Vet Med Sci 63:811–813

Tang Y, Kim M, Carrasco D et al (2005) In vivo assessment of RAS-dependent maintenance of tumor angiogenesis by real-time magnetic resonance imaging. Cancer Res 65:8324–8330

Taskinen M, Karjalainen-Lindsberg ML, Leppa S (2008) Prognostic influence of tumor-infiltrating mast cells in patients with follicular lymphoma treated with rituximab and CHOP. Blood 111:4664–4667

Taskinen M, Jantunen E, Kosma VM et al (2010) Prognostic impact of Cd31-positive microvessel density in follicular lymphoma patients treated with immunochemotherapy. Eur J Cancer 46:2506–2512

Taylor S, Folkman J (1982) Protamine is an inhibitor of angiogenesis. Nature 297:307–312

Tchougounova E, Lundequist A, Fajardo I et al (2005) A key role for mast cell chymase in the activation of pro-matrix metalloprotease-9 and pro-matrix metalloprotease-2. J Biol Chem 280:9291–9296

Textor B, Licht AH, Tuckermann JP et al (2007) JunB required for IgE-mediated degranulation and cytokine release of mast cells. J Immunol 179:6873–6880

Thabrew H, Cairns JA, Walls AF (1996) Mast cell tryptase is a growth factor for human airway smooth muscle. J Allergy Clin Immunol 97:969

Thakurdas SM, Melicoff E, Sansores-Garcia L et al (2007) The mast cell restricted tryptase mMCP-6 has a critical immunoprotective role in bacterial infection. J Biol Chem 282:20809–20815

Theoharides TC, Conti P (2004) Mast cells: the Jekyll and Hyde of tumor growth. Trends Immunol 25:235–241

Theoharides TC, Konstantinidou AD (2007) Corticotropin-releasing hormone and the blood-brain-barrier. Front Biosci 12:1615–1628

Theoharides TC, Kempuraj D, Tagen M et al (2007) Differential release of mast cells mediators and the pathogenesis of inflammation. Immunol Rev 217:65–78

Theoharides TC, Kempuraj D, Marchand J et al (2009) Urticaria pigmentosa associated with acute stress and lesional skin mast-cell expression of CRF-R1. Clin Exp Dermatol 34:163–166

Thompson WD, Campbell R, Evans T (1995) Fibrin degradation and angiogenesis: quantitative analysis of the angiogenic response in the chick chorioallantoic membrane. J Pathol 145:27–37

Thorense S, Tangen M, Hartveit F (1982) Mast cells in the axillary nodes of breast cancer patients. Diagn Histopathol 5:65–67

Thorton SC, Mueller SM, Levine EM (1983) Human endothelial cells: use of heparin in cloning and long term cultivation. Science 222:623–625

Tjabringa GS, Aarbiou J, Ninaber DK et al (2003) The antimicrobial peptide LL-37 activates innate immunity at the airway epithelial surface by transactivation of the epidermal growth factor receptor. J Immunol 171:6690–6696

Tomita M, Matsuzaki Y, Onitsuka T (2000) Effect of mast cell on tumor angiogenesis in lung cancer. Ann Thorac Surg 69:1686–1690

Tomita M, Matsuzaki Y, Edagawa M et al (2001) Association of mast cells and tumor angiogenesis in esophageal squamous cell carcinoma. Dis Esophagus 14:135–138

Toth T, Toth-Jakatics R, Jimi S et al (2000) Cutaneous malignant melanoma: correlation between neovascularization and peritumor accumulation of mast cells overexpressing vascular endothelial growth factor. Hum Pathol 31:955–960

Tournilhac O, Santos DD, Xu L et al (2006) Mast cells in Waldenström's macroglobulinemia support lymphoplasmacytic cell growth through CD154/CD40 signaling. Ann Oncol 17:1275–1282

Travis WD, Li CY, Yam LT et al (1988) Significance of systemic mast cell disease with associated hematologic disorders. Cancer 62:965–972

Tremaine WJ, Brzezinski A, Katz JA et al (2002) Treatment of mildly to moderately active ulcerative colitis with a tryptase inhibitor (APC 2059): an open-label pilot study. Aliment Pharmacol Ther 16:407–413

Tripodo C, Gri G, Piccoluga PP et al (2010) Mast cells and Th 17 cells contribute to the lymphoma-associated pro-inflammatory microenvironment of angioimmunoblastic T-cell lymphoma. Am J Pathol 177:792–802

Triverdi NN, Caughey GH (2010) Mast cell peptidases. Chameleons of innate immunity and host defense. Am J Respir Cell Mol Biol 42:257–267

Tsai M, Shih LS, Newlands GF et al (1991) The rat c-kit ligand, stem cell factor, induces the development of connective tissue-type and mucosal mast cells in vivo. Analysis by anatomical distribution, histochemistry and protease phenotype. J Exp Med 174:125–131

Tsuruda T, Kato J, Hatakeyama K et al (2006) Adrenomedullin in mast cells of abdominal aortic aneurysm. Cardiovasc Res 70:158–184

Tuna B, Yorukoglu K, Unlu M et al (2006) Association of mast cells with microvessel density in renal cell carcinoma. Eur J Urol 50:530–534

Turner GA (1982) Surface properties of the metastatic cell. Invasion Metastasis 2:197–216

Turner HE, Nagy Z, Gatter KC et al (2000) Angiogenesis in pituitary adenomas and in the normal pituitary gland. J Clin Endocrinol Metab 85:1159–1162

Turner RT, Iwaniec UT, Marley K et al (2010) The role of mast cells in parathyroid bone disease. J Bone Miner Res 25:1637–1649

Udagawa T, Fernandez A, Achilles EG et al (2002) Persistence of microscopic human cancers in mice: alterations in the angiogenic balance accompanies loss of tumor dormancy. FASEB J 16:1361–1370

Ullrich SE, Nghiem DX, Khaskina P (2007) Suppression of an established immune response by UVA—a critical role for mast cells. Photochem Photobiol 83:1095–1100

Unna PG (1896) The histopathology of the diseases of the skin. Macmillan, Edinburgh

Uratani Y, Takiguchi-Hayashi K, Miyasaka N et al (2000) Latexin, a carboxypeptidase A inhibitor, is expressed in rat peritoneal mast cells and is associated with granular structures distinct from secretory granules and lysosomes. Biochem J 346:817–826

Utrera-Barillas D, Castro-Manrreza M, Castellanos E et al (2010) The role of macrophages and mast cells in lymphangiogenesis and angiogenesis in cervical carcinogenesis. Exp Mol Pathol 89:190–196

Vacca A, Ribatti D (2006) Bone marrow angiogenesis in multiple myeloma. Leukemia 20:193–199

Vajkoczy P, Farhadi M, Gaumann A et al (2002) Microtumor growth initiates angiogenic sprouting with simulatenous expression of VEGF, VEGF receptor-2, and angiopoietin-2. J Clin Invest 109:777–785

Valent P, Bettelheim P (1992) Cell surface structures on human basophils and mast cells: biochemical and functional characterization. Avd Immunol 52:324–333
van Beijnum J, Dings RP, van der Linden E et al (2006) Gene expression of tumor angiogenesis dissected; specific targeting of colon cancer angiogenic vasculature. Blood 108:2339–2348
Vanderslice P, Ballinger SM, Tam EK et al (1990) Human mast cell tryptase: multiple cDNAs and genes reveal a multigene serine protease family. Proc Natl Acad Sci U S A 87:3811–3815
Varadaradjalou S, Feger F, Thieblemont N et al (2003) Toll-like receptor 2 (TLR2) and TLR4 differentially activate human mast cells. Eur J Immunol 33:899–906
Varayoud J, Ramos JG, Bosquiazzo VL et al (2004) Mast cells degranulation affects angiogenesis in the rat uterine cervix during pregnancy. Reproduction 127:379–387
Veerappan A, Reid AC, Estephan R et al (2008) Mast cell renin and a local renin-angiotensin system in the airway: role in bronchoconstriction. Proc Natl Acad Sci U S A 105:1315–1320
Verbsky JW, McAllister PK, Malone DG (1996) Mast cell activation in human synovium explants by calcium ionophore A23187, compound 48/80, and rabbit IgG anti-human IgE, but not morphine sulfate. Inflamm Res 45:35–41
Venneri MA, De Palma M, Ponzoni M et al (2007) Identification of proangiogenic TIE2-expressing monocytes (TEMs) in human peripheral blood and cancer. Blood 109:5276–5285
Vicaut E, Laemmel E, Stücker O (2000) Impact of serotonin on tumour growth. Ann Med 32:187–194
Vines CM, Prossnitz ER (2004) Mechanisms of G protein-coupled receptor-mediated degranulation. FEMS Microbiol Lett 236:1–6
Vincent AJ, Zhang J, Ostor A et al (2000) Matrix metalloproteinase-1 and-3 and mast cells are present in the endometrium of women using progestin-only contraceptives. Human Reprod 15:123–130
Vliagolftis H, Metcalfe DD (1997) Cell adhesion molecules in mast cell adhesion and migration. In: Bochner BS (ed) Adhesion molecules in allergic diseases. Marcel Dekker, New York, pp 151–172
Vlodavsky I, Korner G, Ishai-Michaeli R et al (1990) Extracellular matrix-resident growth factors and enzymes: possible involvement in tumor metastasis and angiogenesis. Cancer Metastasis Rev 9:203–226
Vlodavsky I, Eldor A, Haimovitz-Friedman A et al (1992) Expression of heparanase by platelets and circulating cells of the immune system: possible involvement in diapedesis and extravasation. Invasion Metastasis 12:112–127
Vogelstein B (1993) The multistep nature of cancer. Trends Genet 9:138–141
von Köckritz-Blickwede M, Goldmann O, Thulin P (2008) Phagocytosis-independent antimicrobial activity of mast cells by means of extracellular trap formation. Blood 111:3070–3080
von Stebut E, Metz M, Milon G et al (2003) Early macrophage influx to sites of cutaneous granuloma formation is dependent on MIP-1alpha/beta released from neutrophils recruited by mast cell-derived TNF-alpha. Blood 101:210–215
Wallon C, Söderholm JD (2009) Corticotropin-releasing hormone and mast cells in the regulation of mucosal barrier function in the human colon. Ann N Y Acad Sci 1165:206–210
Wallon C, Yang PC, Keita AV et al (2008) Corticotropin-releasing hormone (CRH) regulates macromolecular permeability via mast cells in normal human colonic biopsies in vitro. Gut 57:50–58
Walsh LJ, Trinchieri G, Waldorf HA et al (1991) Human dermal mast cells contain and release tumor necrosis factor alpha, which induces endothelial leukocyte adhesion molecule 1. Proc Natl Acad Sci U S A 88:4220–4224
Wang Y, Oliver G (2010) Current views on the function of the lymphatic vasculature in health and disease. Genes Dev 24:2115–2126
Wang XS, Lau HY (2006) Beta-adrenoceptor-mediated inhibition of mediator release from human peripheral blood-derived mast cells. Clin Exp Pharmacol Physiol 33:746–750
Wang JF, Zhang X-F, Groopman JE (2001) Stimulation of beta 1 integrin induces tyrosine phosphorylation of vascular endothelial growth factor receptor-3 and modulates cell migration. J Biol Chem 276:41950–41957

Wang SW, Oh CK, Cho SH et al (2005) Amphiregulin expression in human mast cells and its effect on the primary human lung fibroblasts. J Allergy Clin Immunol 115:287–294
Watnick RS, Cheng YN, Rangarajan A et al (2003) Ras modulates Myc activity to repress thrombospondin-1 expression and increase tumor angiogenesis. Cancer Cell 3:219–231
Weber A, Knop J, Maurer M (2003) Pattern analysis of human cutaneous mast cell populations by total body surface mapping. Br J Dermatol 148:224–228
Weidner N, Sample JP, Welch WR et al (1991) Tumor angiogenesis correlates with metastasis in invasive breast carcinoma. N Engl J Med 324:1–8
Weill P (1919) Mastzellenstudien an Sarkkommetastasen. Folia hematol (Lpz) 23:185–196
Weiss L (1979) Dynamic aspects of cancer cell populations in metastasis. Am J Pathol 97:601–608
Weller K, Foitzik K, Paus R et al (2006) Mast cells are required for normal healing of skin wounds in mice. FASEB J 20:2366–2368
Welsh TJ, Green RH, Richardson D et al (2005) Macrophage and mast cell invasion of tumor cell islets confers a marked survival advantage in non-small-cell lung cancer. J Clin Oncol 23:8959–8567
Wershil BK, Tsai M, Geissler EN et al (1992) The rat c-kit ligand, stem cell factor, induce c-kit receptor-dependent mouse mast cell activation in vivo. Evidence that signaling through the c-kit receptor can induce expression of cellular function. J Exp Med 175:245–255
Westphal E (1891) Über Mastzellen. In: Ehrlich P (ed) Farbenanalytische Untersuchungen, Hirschwald, Berlin, pp 17–41
Whitelock JM, Murdoch AD, Iozzo RV et al (1996) The degradation of human endothelial cell-derived perlecan and release of bound basic fibroblast growth factor by stromelysin, collagenase, plasmin, and heparanases. J Biol Chem 271:10079–10086
Wilks JW, Scott PS, Urla LK et al (1991) Inhibition of angiogenesis with combination treatments of angiostatic steroids and suramin. Int J Radiat Biol 60:73–77
Wilk M, Liszka L, Palen P et al (2010) Intensity of angiogenesis and mast cell infiltration in cervical intraepithelial and invasive lesions. Are they correlated? Pathol Res Pract 206:217–222
Wille J, Kydonieus A, Murphy G (1999) Cis-urocanic acid induces mast cell degranulation and release of preformed TNF-alpha: a possible mechanism linking UVB and cis-urocanic acid to immunosuppression of contact hypersensitivity. Skin Pharmacol Physiol 12:18–27
Williams HU (1900) A critical summary of recent literature on plasma-cells and mast cells. Am J Med Sci 119:702–709
Winandy S, Brown M (2007) No DL1 Notch ligand? GATA be a mast cell. Nature Immunol 8:796–798
Wodnar-Filipowicz A, Heusser CH, Moroni C (1989) Production of the haematopoietic growth factor GM-CSF and interleukin-3 by mast cells in response to IgE receptor-mediated activation. Nature 339:150–152
Wojtecka-Lukasik E, Maslinski S (1992) Fibronectin and fibrinogen degradation products stimulate PMN-leukocytes and mast-cell degranulation. J Physiol Pharmacol 43:173–181
Wright CD, Havill AM, Middleton SC et al (1999) Inhibition of allergen-induced pulmonary responses by the selective tryptase inhibitor 1,5-bis-[4-[(3-carbamimidoyl-benzenesulfonylamino)-methyl]-phenoxy]-pentane (AMG-126737). Biochem Pharmacol 58:1989–1996
Yamazaki S, Yokozechi H, Satoh T et al (1998) TNF-alpha, RANTES, and MCP-1 are major chemoattractants in murine Langerhans cells to the regional lymph nodes. Exp Dermatol 7:35–41
Yancopoulos GD, Davis S, Gale NW et al (2000) Vascular specific growth factors and blood vessels formation. Nature 407:242–248
Yoshioka M, Fukuishi N, Kubo Y et al (2008) Human cathelicidin CAP18/LL-37 changes mast cell function toward innate immunity. Biol Pharm Bull 31:212–216
Young AJ, Seabrook TJ, Marston WL et al (2000) A role for lymphatic endothelium in the sequestration of recirculating gamma delta T cells in TNF-alpha-stimulated lymph nodes. Eur J Immunol 30:327–334
Yayou A, Klagsbrun M, Esko JD et al (1991) Cell surface, heparin-like molecules are required for binding of basic fibroblast growth factor to its high affinity receptor. Cell 64:841–848

Zagzag D, Hooper A, Friedlander DR et al (1999) In situ expression of angiopoietins in astrocytomas identifies angiopoietin-2 as an early marker of tumor angiogenesis. Exp Neurol 159:391–400

Zauberman H, Michaelson IC, Bergmann F et al (1969) Stimulation of neovascularization of the cornea by biogenic amines. Exp Eye Res 8:77–83

Zhang W, Stoica G, Tasca SI et al (2000) Modulation of tumor angiogenesis by stem cell factor. Cancer Res 60:6757–6762

Zhao W, Oskeritzian CA, Pozez AL et al (2005) Cytokine production by skin-derived mast cells: endogenous proteases are responsible for degradation of cytokines. J Immunol 175:2635–2642

Zhou JS, King W, Friend DS et al (2007) Mast cell deficiency in Kit(W-sh) mice does not impair antibody-mediated arthritis. J Exp Med 204:2797–2802

Zidaire E, Martinez A, Garayoa M et al (2006) Adrenomedullin is a cross-talk molecule that regulates tumor and mast cell function during human carcinogenesis. Am J Pathol 168:280–291